The Progress of Equine Medical Research in China and Beyond

Guest Editors

Jing Li

Gayle Leith

MDPI

Basel • Beijing • Wuhan • Barcelona • Belgrade • Novi Sad • Cluj • Manchester

Guest Editors

Jing Li
College of Veterinary
Medicine
China Agricultural University
Beijing
China

Gayle Leith
College of Veterinary
Medicine
University of Arizona
Tucson
USA

Editorial Office
MDPI AG
Grosspeteranlage 5
4052 Basel, Switzerland

This is a reprint of the Special Issue, published open access by the journal *Veterinary Sciences* (ISSN 2306-7381), freely accessible at: https://www.mdpi.com/journal/vetsci/special_issues/OT7J2IJ902.

For citation purposes, cite each article independently as indicated on the article page online and as indicated below:

Lastname, A.A.; Lastname, B.B. Article Title. *Journal Name* **Year**, *Volume Number*, Page Range.

ISBN 978-3-7258-4691-7 (Hbk)
ISBN 978-3-7258-4692-4 (PDF)
https://doi.org/10.3390/books978-3-7258-4692-4

Contents

veterinary sciences

MDPI

Case Report

Pseudomonas-Enterobacter Co-Infection Drives Cellulitis and Lymphangitis in Equines: A Case Report

Xiangning Huang [1,2], Renjie Deng [1], Haoen Huang [1], Huisheng Xie [3] and Aolei Chen [1,2,*]

[1] College of Veterinary Medicine, South China Agricultural University, 483 Wushanlu, Guangzhou 510640, China; nelliehuangxn@stu.scau.edu.cn (X.H.)
[2] Guangdong Technological Engineering Research Center for Pet, 483 Wushanlu, Guangzhou 510640, China
[3] Chi University, 9650 West Highway 318, Reddick, FL 32686, USA; shen@chiu.edu
* Correspondence: chenolay@scau.edu.cn

Simple Summary: Cellulitis/lymphangitis is a common disease in horses. Clinicians diagnose these diseases based on empirical judgement, namely the clinical presentation such as fever, limb swelling and acute lameness. This may lead to misdiagnoses or missed diagnoses or even practice against antibiotic stewardship. However, there are limited published reports focusing on diagnostic tests and integrative treatment regarding cellulitis/lymphangitis. This case report first documented a comprehensive approach using bacterial culture, bacterial genomics analysis and lymphangiograms to confirm the diagnosis.

Abstract: This case report detailed a rare co-infection of *Pseudomonas asiatica* and *Enterobacter hormaechei* in a 9-year-old warmblood mare, leading to severe cellulitis and secondary lymphangitis following traditional hoof blood-letting therapy. The mare exhibited acute limb swelling, fever, cutaneous ulceration, lymphatic dysfunction and unknown anemia. Comprehensive diagnostics, including bacterial culture, whole-genome sequencing, anti-elastin antibody (AEAb) ELISA, and diagnostic imaging, confirmed the pathogens causing cellulitis and secondary lymphangitis. AEAb levels were elevated, correlating with lymphatic degradation, while radiography and lymphangiography ruled out laminitis but identified tortuous lymphatic vessels. The treatment integrated systemic antimicrobials, anti-inflammatory therapy, combined decongestive therapy, and traditional Chinese herbal medicine, resulting in resolution of infection, improved hematological parameters, and restored athletic performance. The therapeutic regimen primarily included gentamicin, enrofloxacin, oxytetracycline, and the Wei Qi Booster. The case highlights the critical role of pathogen-directed antimicrobial selection and the potential benefits of combining conventional and holistic therapies. This report emphasizes the necessity of early, multifaceted interventions to prevent life-threatening complications in equine cellulitis–lymphangitis cases.

Keywords: horse; lymphocutaneous infections; *Pseudomonas asiatica*; *Enterobacter hormaechei*; traditional Chinese veterinary medicine (TCVM)

1. Introduction

Cellulitis is an acute infection of the skin and subcutaneous tissues that is typically caused by bacteria. In most cases we cannot identify the bacterial pathogens, with many negative cultural results turning out to be due to the previous use of antibiotics [1]. Epidemiologically, cellulitis represents a common condition, affecting individuals across various age groups, and its incidence is particularly high among leisure horses and sports horses [2]. Although most cases respond well to appropriate antimicrobial therapy, complications—including sepsis, abscess and secondary lymphangitis—can adversely affect the prognosis. Poor prognosis may lead to motor dysfunction and, worse still, euthanasia [3–5]. On the other hand, lymphangitis is inflammation and/or infection of the lymphatic vessels [6]. Clinically, lymphangitis shares a similar presentation to cellulitis, which sometimes makes these two words interchangeable.

Common clinical signs of cellulitis include localized erythema, swelling, pain, and fever [3,4]. However, these features broaden the clinical spectrum of the disease and emphasize the need for careful and comprehensive evaluation. Currently, the clinical diagnosis of cellulitis basically relies on the evaluation of characteristic symptoms and the clinician's empirical judgment [4,7]. Therefore, reliance on subjective assessment can lead to misdiagnoses or missed diagnoses.

The standard treatment regimen for cellulitis typically includes a combination of systemic antibiotic therapy, supporting treatment, and in severe cases, incision and drainage [8,9]. According to the study of Adam and Southwood, pain management is also an important factor that affects the therapeutic outcome [10]. However, various severe complications such as avulsion of the hoof capsule, dermal necrosis, laminitis, and widespread thrombosis and permanent lameness may lead to a decrease in the survival rate [10–12]. Therefore, timely diagnosis, comprehensive evaluation, early intervention and tailor-made treatment plans are essential to improve prognosis.

This case describes the clinical manifestation, diagnosis and treatment of lymphangitis secondary to cellulitis in a 9-year-old warmblood mare caused by *Pseudomonas asiatica* and *Enterobacter hormaechei* coinfection. This case report highlights the multimodal diagnostic methods and the support given to the animal in a holistic manner.

2. Materials and Methods

2.1. Case Description

A 9-year-old warmblood mare from an equestrian club, used for show jumping, developed swelling in the right hind limb and a high fever after traditional hoof bloodletting therapy treatment a day before presentation. This is an ancient method where an iron nail was inserted into a specific acupoint—Hou-ti-tou—to relieve inflammation and pressure. We believed improper handling was a source of infection.

An initial physical examination revealed a high fever at 40.4 °C and the swelling gradually extended from the pastern to above the stifle joint. Pitted edema in the affected limb was noted. The skin of the affected limb became thin, erythematous, and ulcerated, and began oozing yellow exudate (Figure 1A). The coronary band was swollen, with excessive skin thickening and mild hoof-wall detachment (Figure 1B). In later stages, small pustules appeared on the skin of the gaskin and ruptured (Figure 1C).

An integrative medical treatment plan was implemented, including antimicrobial therapy and regional limb perfusion, anti-inflammation therapy, manual massage, combined decongestive therapy (CDT), and exercise rehabilitation to prevent secondary laminitis. Qi-Blood are the fundamental substances that maintain life activities and body functions in TCVM theory. The Chinese herbal medicine Wei Qi Booster was administered to tonify the Qi-Blood and strengthen immune functions, as well as the Heat for enhancing the antibiotic effects and accelerating the healing process. Gentamicin (6.6 mg/kg, I.V. q24h) and enrofloxacin (5 mg/kg, I.V. q24h) were given for 2 weeks with compression bandages and later changed to oxytetracycline (5 mg/kg, I.V. q12h) for 2 weeks when exudate recurred. Phenylbutazone (4.4 mg/kg, I.V. q12h on day 1 and decreased to 2.2 mg/kg, I.V. q12h) was prescribed to decrease inflammation and relieve pain. Manual massage was intended to help lymphatic return. In addition, the Chinese herbal medicine Wei Qi Booster (45 g, PO q12h for 10 d and 15 g, PO, q12h for 20 d), was orally given. This formula primarily contains codonopsis root (Dang Shen), astragalus root (Huang Qi), dong quai root (Dang Gui), and so on. In our 3-month follow-up, the mare was reported to be in good condition with healed skin and recovered athletic performance. In addition, the body condition score (BCS) of the mare had gone from 3/9 to 5/9.

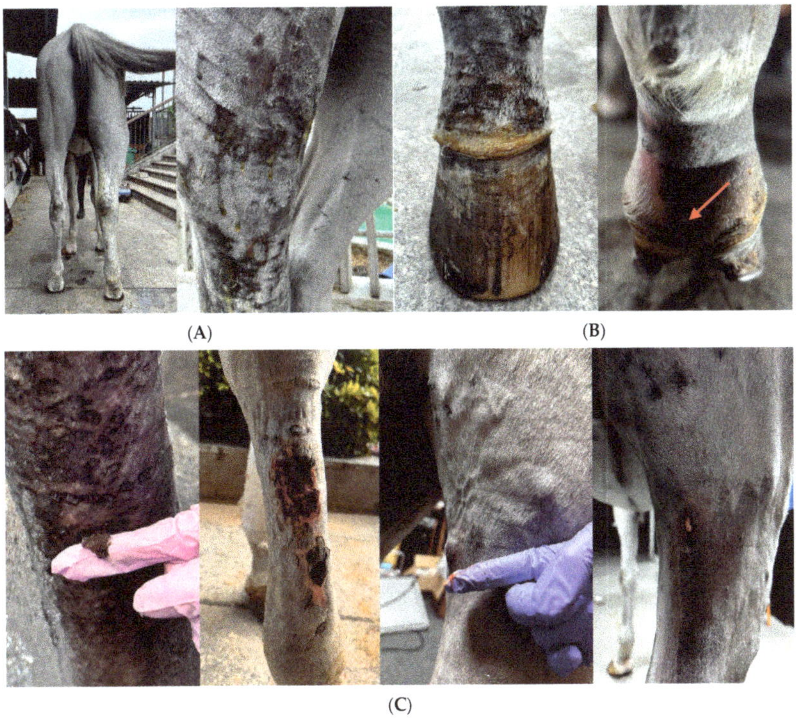

Figure 1. Clinical manifestation. (**A**) Acute swelling of the distal limb of the right hind leg, extending from the pastern to above the stifle joint with oozing straw-colored exudate. (**B**) Swollen coronary band of the right hind limb with mildly detached hoof wall at the plantar aspect (red arrow). (**C**) Skin sloughing, skin necrosis and ruptured pustules.

2.2. Blood Examination

Blood samples were collected from the jugular vein for complete blood count (CBC) (BC-500, Mindray, Shenzhen, China), chemistry panel (Nx700i, Fujifilm, Iwate, Japan), serum amyloid A (SAA) (Vcheck-V200, Bionote, Gyeonggi-do, Republic of Korea) evaluation at South China Agricultural University Veterinary Teaching Hospital (Guangzhou, China).

2.3. Bacteriologic Examination

Two per cent chlorhexidine was used to clean and sterilize the skin to avoid false positives from the normal skin flora. A linear array probe was gently applied on the lateral tarsal joint region for scanning until we found the anechoic region. A 0.7 × 24 mm butterfly needle was inserted into the fluid pocket and confirmed with the ultrasound. Straw-colored exudate was collected for bacterial culture. Samples were sent to Zoetis Reference Laboratory (Shanghai, China) for bacterial culture and antimicrobial susceptibility tests using the VITEK 2 COMPACT system and the Kirby–Bauer disc diffusion method [13].

2.4. Bacterial Genomics Analysis

Sample 1 and Sample 2 were bacterial DNA isolated from bacterial culture. Bacterial DNA was extracted using MagPure Bacterial DNA Kit (D6361-02, Magen, Guangzhou, China). DNA concentration was determined via Qubit4.0 (Thermo, Q33226, Budapest, Hungary). DNA integrity was assessed by 1% agarose gel electrophoresis. The whole genome DNA was randomly fragmented to an average size of 200–400 bp. The selected fragments were put through end-repair, 3′ adenylated, adapter ligation, and PCR amplifying. After purification with the magnetic beads, the library was qualified by the Qubit 4.0 fluorometer and the length of library was assessed by 2% agarose gel electrophoresis. The qualified libraries were sequenced on the Illumina NovaSeq 6000 platform at Sangon Biotech (Shanghai, China). After sequencing, raw reads were filtered via Trimmomatic (v0.36) by removing adaptors and low-quality reads, then clean reads were obtained. Genome assembly was carried out using SPAdes (v3.15) and the Gapfiller (v1.11) was used for filling gaps. Gene predictions were generated using the National Center for Biotechnology Information (NCBI)nr database.

2.5. Anti-Elastin Antibody (AEAb) ELISA

Blood samples were obtained through the jugular vein from the animals and naturally coagulated at room temperature for 20 min. Samples were centrifugedat 2–8 °C for 20 min (2000 rpm) and the sera were carefully collected. The samples were stored at −80 °C for ELISA assay. Sera AEAb were tested by Horse ELN Ab ELISA kit (MeiMian Industrial, Jiangsu, China). The detective range for AEAbs was 10–160 ng/L. The procedure involved preparing a standard curve through sequential dilutions of the original standard (160 mg/L) to generate concentrations ranging from 10 to 160 mg/L. Samples were diluted 5-fold and added to microplate wells alongside blank controls. After incubation at 37 °C for 30 min, the plate was washed repeatedly to remove unbounded components. Horseradish peroxidase (HRP)-conjugate reagent was introduced to all wells except the blanks, followed by another incubation and washing cycle. Color development was initiated by adding chromogen substrates under light-protected conditions at 37 °C for 10 min. The reaction was terminated using a stop solution, and absorbance was measured at 450 nm within 15 min, with blank wells serving as the reference. All serum samples were repeated three times as technical replicates.

2.6. Survey Radiograph, Lymphangiogram and Ultrasonography

The animal was sedated with xylazine (0.3 mg/kg). The skin was scrubbed with 2% chlorhexidine. **Survey radiograph**: Lateral and craniocaudal projections were taken in a setting of 90 kVp, 1 mA/s. **Lymphangiogram**: 10 mL of iodinated contrast medium (Iohexol, 17.5 g(I):50 mL, Cisen-Pharma, Jining, China) was injected subcutaneously at the level of the plantar aspect of the distal pastern followed by 2 mL of lidocaine (20 mg/mL) for local anesthesia. Lateral radiographs (90 kVp, 1 mA/s) of the distal extremities were recorded 10 min after the injection and manual massage (Figure 2). However, due to difficult manipulation of the animal, only 2 mL of contrast medium was successfully injected. **Ultrasonography:** a brief ultrasound examination of the distal limb using a Vetus E7pro ultrasound machine (Mindray Animal Medical Technology, Shenzhen, China) with a 7–13 MHz linear array transducer.

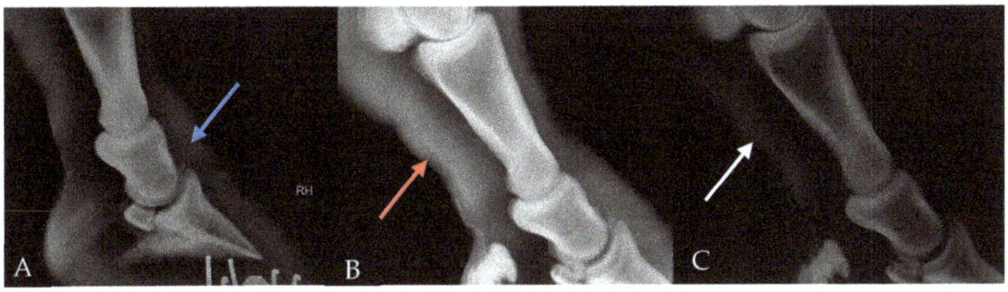

Figure 2. Survey radiograph and lymphangiogram of the right hind limb. (**A**) Survey radiograph of right foot, showing a bony fragment of proximal P3 (blue arrow). (**B**) Lymphangiogram showed multiple tortuous tubular structures with one specifically dilated (red arrow). (**C**) Increased contrast of B to highlight the dilated structure (white arrow).

2.7. Data Analysis

AEAb levels were expressed as optical densities (OD's). Relative expression levels are presented as mean ± standard deviation (SD). The statistical significance of differences was evaluated with the Welch's t-test using GraphPad Prism v10.0.1 software (* p <0.05).

3. Results

3.1. Blood Work Results

Laboratory results showed inflammatory leukogram and mild regenerative anemia initially. During the treatment period, leukocytosis was greatly controlled while the anemia still existed for unknown reasons. Serum amyloid A (SAA) declined to normal indicating the acute inflammation period had passed. At the same time, serum biochemistry showed normal liver and renal function while there was a decrease in the A/G ratio (0.5; ref. 1.5–25). The decrease in serum albumin may have contributed to the excessive lymphatic effusion. Ten days after the use of Wei Qi Booster, the anemia was resolved and the HCT level maintained at 36.3% 2 months after finishing the medication (Table 1).

Table 1. Blood work results.

	10.25	10.27	10.30	11.02	11.16	12.23	3.15	Units	Reference Range
RBC	5.68	5.49	6.65	5.9	5.59	9.65	7.92	10^{12}/L	5.3–10.5
HGB	88 ↓	84 ↓	101	89 ↓	86 ↓	155	131	g/L	100–170
HCT	24.2 ↓	21.7 ↓	26.7 ↓	23.6 ↓	22.9 ↓	40.1	36.3	%	30–49
WBC	22.24 ↑	19.11 ↑	11.24	13.18 ↑	7.47	10.70	7.79	10^9/L	5.00–12.00
Lymphocytes	1.27 ↓	1.8	2.39	2.16	2.70	3.25	3.00	10^9/L	1.32–5.86
Neutrophils	20.15 ↑	15.78	8.40 ↑	10.44 ↑	4.02	6.73	4.24	10^9/L	2.18–6.96
Monocytes	0.81	1.32 ↑	0.35	0.50	0.38	0.37	0.31	10^9/L	0.05–0.92
NEUT%	90.6 ↑	82.6 ↑	74.8 ↑	79.3 ↑	53.8	62.9	54.4	%	38.0–70.0
LYMPH%	5.8 ↓	9.4 ↓	21.3 ↓	16.4 ↓	36.1	30.4	38.5	%	25.0–62.0
Serum amyloid A(SAA)	/	/	101.2	/	<5	/	/	Ug/mL	<10
Albumin	/	/	/	24.0 ↓	23.0 ↓	30 ↓	/	g/L	40–55
Globulin	/	/	/	53.0 ↑	46.0 ↑	49 ↑	/	g/L	20–45
Alb/Glob	/	/	/	0.5 ↓	0.5 ↓	0.6 ↓	/	g/L	1.5–2.5

Remark: "↑" denotes values above the upper limit of the reference range; "↓" denotes values below the lower limit of the reference range. "/" means test was not performed.

3.2. Diagnostic Imaging Results

An X-ray was performed to rule out other musculoskeletal diseases that could have caused acute swelling in the right hind limb. The lateral view of the survey radiograph showed soft tissue swelling and normal bone structures except for a bony fragment of proximal distal phalanx (P3), which previously existed and caused no lameness. No signs indicated potential secondary laminitis.

A lymphangiogram revealed several dilated tortuous lymphatic vessels in the plantar aspect of the distal limb (Figure 2). Ultrasound examination revealed marked subcutaneous and fascial oedema. We also noticed a hyperechoic subcutaneous tissue layer and cobblestone pattern at the coronary band indicating cellulitis.

3.3. Bacteriologic Examination and Pathogen Identification

Two bacterial genera, *Pseudomonas* and *Enterobacter* (phylogenetic analysis is attached in the (Figures S1 and S2), were identified as Gram-negative bacteria which are opportunistic pathogens. Both showed resistance to penicillin, cephalosporin, sulfamethoxazole, and chloramphenicol but were sensitive to gentamicin. Surprisingly, both bacteria were intermediate to enrofloxacin (Table S1). Results from bacterial genomics analysis further confirmed the pathogen species as *Pseudomonas asiatica* and *Enterobacter hormaechei*. Genomics data also revealed several multidrug-resistant genes such as *mdtE, mdtA, mdtK, mdtB, stp*, etc. This may explain why these pathogens were resistant to multiple antimicrobial categories.

3.4. Anti-Elastin Antibody (AEAb) ELISA Results

A total of twelve horses were sampled and analyzed, and the data were divided into the following cohorts: Group 1: the AEAb baseline level of the animal, sampled 39 days after onset; Group 2: AEAb level of the animal, re-sampled at 59 days; Group 3: AEAb level of healthy warmblood controls; Group 4: AEAb level of asymptomatic warmblood horses with chronic lymphedema; Group 5: AEAb level of healthy draft horse controls.

There was a statistically significant difference ($p < 0.05$) between Group 1 and Group 2, indicating the AEAb level may have increased as the disease progressed (Table 2); Group 5 showed statistically significant differences compared to Group 3 ($p < 0.05$), indicating the AEAb level may have breed difference (Table 3). Group 4 had statistically significant differences compared to Group 3 ($p < 0.05$) combined with the fact that the mean AEAb level of the animals (Group 1 and Group 2) is higher than warmblood horses with chronic lymphedema (Group 4), indicating that the AEAb level increased in affected horses.

Table 2. Mean ± SD optical density (OD) and median of antibodies in the affected horse.

	Group 1 Baseline	Group 2 Follow-Up
Number of horses	1	1
Mean OD ± SD	0.530 ± 0.073	0.675 ± 0.106 *
Median	0.569	0.704

* Statistically significant ($p < 0.05$), compared to baseline.

Table 3. Mean ± SD optical density (OD) and median of antibodies in different horse breeds.

	Group 3 Healthy Warmblood Horses	Group 4 Lymphedematous Warmblood Horses	Group 5 Healthy Draft Horses
Number of horses	4	4	4
Mean OD ± SD	0.332 ± 0.041	0.436 ± 0.044 *	0.515 ± 0.014 *
Median	0.325	0.420	0.516

* Statistically significant ($p < 0.05$), compared to healthy warmblood.

4. Discussion

Cellulitis is a common disease in horses. It can be caused by a tiny penetration and, without proper management, can lead to severe complications like endocarditis, osteomyelitis, and toxic shock [14]. Therefore, timely diagnostic examinations and early therapeutic interventions are crucial to prevent serious complications. Cellulitis can present with a wide range of symptoms, including swelling, localized fever, pain, and oozing serum. Some clinical signs are common and may be confused with other diseases, while some clinical signs are rare. A previous case report described hindlimb cellulitis that triggered lymphangitis, with bacterial cultures yielding *Staphylococcus aureus* and *Escherichia coli*. In the later stages of the disease, detachment of the hoof capsule occurred, ultimately resulting in euthanasia of the horse [11]. This case report documented some uncommon symptoms of cellulitis such as an extensive lymphedema region along with pustules, skin sloughing, and skin ulceration, and unknown anemia. From a TCVM perspective, Qi serves as the body's primary defensive mechanism, preventing the invasion of pathogens, including bacteria. When Qi is deficient, this protective barrier is weakened, allowing bacteria to enter the body more easily and cause infections. TCVM also believes that the Blood is generated from Qi. Therefore, insufficient Qi can impair the Blood production, ultimately leading to anemia.

P. asiatica and *E. hormaechei* are both opportunistic pathogens that can inhabit wet environments including water and soil [15,16]. *P. asiatica* is a newly identified species firstly isolated in Japan that seldom causes severe infections unless the host went through invasive procedures or was immunocompromised [17]. No disease caused by *P. asiatica* in horses has been reported to date, which may be partially attributed to the lack of bacterial identification in clinical practice. *E. hormaechei* is a member of the Enterobacter cloacae complex (ECC) which often leads to severe nosocomial infections through its multidrug resistance. Research showed that *E. hormaechei* is more toxic compared to other ECC members, and can cause cellulitis as well as pneumonia, urinary infections, and so on [18]. They both share an efflux pump genes mechanism, along with the activity of β-lactamase production, jointly contributingto their multidrug resistance. It is worth noting that both strains have been implicated in nosocomial outbreaks in human hospitals [15,19,20]. In 2022, researchers in China isolated *Pseudomonas putida* strains, a strain close to *P. asiatica*, harboring five distinct β-lactamases, highlighting the escalating threat of antibiotic-resistant bacteria as a critical public health issue spanning both veterinary and human medicine [21].

Chronic progressive lymphedema (CPL) is thought to be a genetic disease that commonly affects draft horses and shares a similar clinical presentation with cellulitis. This disease is characterized by progressive swelling of the distal limbs, accompanied by scaling, marked dermal fibrosis, and the development of skinfolds and nodules, which are often complicated by secondary infections [22]. Under pathological circumstances elastin is increasingly degraded into peptides (elastin derived peptides: EPs), which are released into the circulation. The immune system initiates an immune response against the released peptides, generating anti-elastin antibodies (AEAb) [23]. Elastin is a major supportive component of lymphatics, and thus essential for their function, while elevated levels of AEAb suggest the occurrence of lymphatic pathology [24]. This study revealed significantly elevated AEAb levels in a warmblood horse affected by lymphangitis. Although the clinical signs improved, the animal's AEAb level kept increasing. This may be due to a 70-year half-life of AEAbs [25]. As a result, we conclude that there is a breed-specific difference in AEAb levels, which is inconsistent with what Van Brantegem proposed [26]. Risk factors such as age, gender, pregnancy status, and laboratory differences can affect AEAb levels [27]. Previous studies have explored whether the ELISA test can be used as an indicator to diagnose CPL. A study revealed that the ELISA technique used to identify serum levels of AEAb was a valuable tool for the diagnosis of CPL and associated with the severity of the disease, while another study showed that ELISA procedure was not useful for CPL detection due to low sensitivity [26,27]. Brys M proposed a multimodal-integrated diagnostic approach to confirm CPL [22]. Although our animal exhibited elevated AEAb levels, she did not completely fit into the diagnostic criteria for CPL. According to our data, we believe that AEAb levels can be used as a conjunctive test for lymphangiopathy. However, the disadvantage of this study was the lack of a sufficient number of samples. Further research is needed on whether AEAbs can be used as an indicator of lymphangitis.

In this case, a holistic treatment approach was adopted, including antimicrobial and anti-inflammatory therapy, and traditional Chinese medicine, pain control, and CDT. CDT is a treatment for human lymphoedema. It can be an effective treatment for limbs swelling in horses which consists of phase I and phase II. Phase I involves daily manual lymph drainage (MLD), skin care, multi-layer short-stretch bandaging, and exercise to rapidly reduce edema and soften fibrotic tissue until the limb volume stabilizes. Phase II focuses on maintenance through continued skin care, exercise, and a transition to elastic compression garments, with MLD tapered to minimal frequency, ensuring long-term prevention of fluid re-accumulation [28]. Pain control was of paramount importance during treatment, and exercise played a critical role in recovery. Routine care included mild washing after exercise—strictly avoiding irritation and refraining from vigorous scrubbing. Considering the risk of laminitis and hoof capsule detachment, an X-ray was implemented and the result showed no evidence of laminitis. With strategic management, we resolved the potential severe complications and hoof capsule detachment, and the animal showed a good condition by a 5-month follow-up (Figure 3).

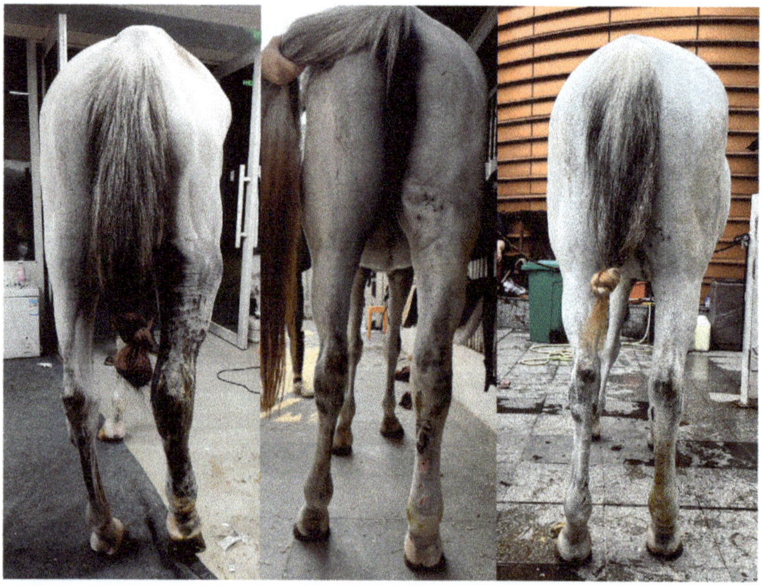

Figure 3. Clinical improvement. From left to right: initial presentation, 6-week follow-up, 5-month follow-up.

In the later phase of treatment, the traditional Chinese medicine Wei Qi Booster was used to address persistent unknown anemia and lymphedema. In TCVM, Qi is responsible for the movement of body fluids. When Qi is deficient, this physiological function is impaired, leading to the accumulation of fluids, which may manifest as edema, including lymphedema. Qi also plays a vital role in the generation of the Blood, which nourishes all tissues, including the skin. Therefore, tonifying Qi can support skin healing and overall tissue regeneration. Exploring the appropriate use of traditional Chinese medicine may be an adjunct therapy for cellulitis and lymphangitis in the future.

Although our multidimensional data provides a diagnostic reference framework for similar cases, several limitations persist. First, the application of diagnostic imaging examinations, especially the ultrasound scan, is highly dependent on the skill level of the operator. We should also consider the availability of advanced equipment in clinical settings. Additionally, the prohibitively increased cost of comprehensive diagnosis must be acknowledged. To address these challenges, we recommend prioritizing bacterial isolation and identification alongside antimicrobial susceptibility testing to facilitate timely initiation of antibiotic therapy. Subsequent diagnostic steps, such as diagnostic imaging or an AEAb ELISA test, should be tailored to the patient's clinical progression and the owner's financial capacity, ensuring a balance between diagnostic rigor and practical feasibility. While the herbal formula used in this case demonstrated efficacy in the affected animal, its mechanisms of action still lack support in the current literature. When applied to different clinical presentations, a thorough TCVM diagnostic evaluation should precede the selection of herbal therapies to ensure treatment alignment with specific pathological manifestations.

5. Conclusions

This case report presents a 9-year-old warmblood mare diagnosed with lymphangitis secondary to cellulitis caused by a rare *Pseudomonas asiatica* and *Enterobacter hormaechei* coinfection. This report highlights that comprehensive diagnostic workups must precede empirical treatment to prevent life-threatening complications like hoof capsule detachment. Notably, combining antibiotic treatment with traditional Chinese herbal therapy, pain management, and CDT can achieve a better prognosis.

Supplementary Materials: The following supporting information can be downloaded at: https://www.mdpi.com/article/10.3390/vetsci12060574/s1, Figure S1: Phylogenetic analysis of *P. asiatica*; Figure S2: Phylogenetic analysis of *E. Hormaechei*; Table S1: Antimicrobial susceptibility test results.

Author Contributions: Conceptualization, X.H. and A.C.; methodology, X.H. and A.C.; software, X.H. and R.D.; validation, X.H. and A.C.; investigation, A.C., X.H., R.D. and H.H.; clinical case management, A.C., R.D., H.H. and X.H.; resources, A.C.; data curation, A.C., X.H. and R.D.; writing—original draft preparation, X.H.; writing—review and editing, A.C., X.H., R.D., H.X. and H.H.; visualization, X.H. and A.C.; supervision, A.C. All authors have read and agreed to the published version of the manuscript.

Funding: This study was supported by the National Natural Science Foundation of China (32302940).

Institutional Review Board Statement: Not applicable.

Informed Consent Statement: Not applicable.

Data Availability Statement: The original contributions presented in this study are included in the article. Further inquiries can be directed at the corresponding author.

Acknowledgments: The authors would like to acknowledge the assistance from Liang Deng of Shenyang Agricultural University for plasma collection of draft horses in Northeast China. We would also like to thank Huisheng Xie from Chi University for his input on Traditional Chinese Veterinary Medicine in this case.

Conflicts of Interest: The authors declare no conflicts of interest.

References

1. Cooper, H.E.; Davidson, E.J.; Slack, J.; Ortved, K.F. Treatment and Outcome of Eight Horses with Limb Cellulitis and Septic Tendonitis or Desmitis. *Vet. Surg.* **2021**, *50*, 1542–1552. [CrossRef] [PubMed]
2. Braid, H.R.; Ireland, J.L. A Cross-sectional Survey of the Diagnosis and Treatment of Distal Limb Cellulitis in Horses by Veterinary Surgeons in the United Kingdom. *Equine Vet. Educ.* **2022**, *34*, E234–E244. [CrossRef]
3. Adam, E.N.; Southwood, L.L. Primary and Secondary Limb Cellulitis in Horses: 44 Cases (2000–2006). *J. Am. Veter-Med Assoc.* **2007**, *231*, 1696–1703. [CrossRef]
4. Rendle, D. Cellulitis and Lymphangitis. *Medicine* **2017**, *1*, 16–20. [CrossRef]
5. Oddsdóttir, C.; Sigurðardóttir, Ó.G.; Friðriksdóttir, V.; Svansson, V.; Bragason, B.Þ.; Björnsdóttir, S. Severe Subcutaneous Infection with *Clostridium septicum* in a Herd of Native Icelandic Horses. *Acta Vet. Scand.* **2025**, *67*, 8. [CrossRef]
6. Kano, Y.; Momose, T. Acute Lymphangitis. *Clevel. Clin. J. Med.* **2020**, *87*, 129–130. [CrossRef] [PubMed]
7. Braid, H. Diagnosing and treating cellulitis in horses. *InPractice* **2024**, *46*, 536–540. [CrossRef]
8. Adam, E.N. Cellulitis: Any Change? *Equine Vet. Educ.* **2019**, *31*, 625–626. [CrossRef]
9. Long, B.; Gottlieb, M. Diagnosis and Management of Cellulitis and Abscess in the Emergency Department Setting: An Evidence-Based Review. *J. Emerg. Med.* **2022**, *62*, 16–27. [CrossRef]
10. Adam, E.N.; Southwood, L.L. Surgical and Traumatic Wound Infections, Cellulitis, and Myositis in Horses. *Vet. Clin. N. Am. Equine Pract.* **2006**, *22*, 335–361. [CrossRef]
11. Duggan, M.; Mair, T.; Fews, D. Equine Limb Cellulitis/Lymphangitis Resulting in Distal Limb Ischaemia and Avulsion of the Hoof Capsule. *Equine Vet. Educ.* **2022**, *34*, E406–E412. [CrossRef]
12. Fjordbakk, C.T.; Arroyo, L.G.; Hewson, J. Retrospective Study of the Clinical Features of Limb Cellulitis in 63 Horses. *Vet. Rec.* **2008**, *162*, 233–236. [CrossRef] [PubMed]
13. Liu, Y.; Zhai, Y.; Jiang, C.; Liu, H.; Li, Z.; Yuan, Y.; Song, J.; Yuan, S. Surveillance of Antimicrobial Resistance in Hospitalized Companion Animals in China in 2022–2023. *JAC-Antimicrob. Resist.* **2024**, *7*, dlaf007. [CrossRef] [PubMed]
14. Reis, J.; Carmo, F.; Soares, I.; Salvado, C.; Fidalgo, M. Pseudomonas Mendocina Bacteraemia Secondary to Cellulitis—A Report and Brief Series of Cases. *Eur. J. Case Rep. Intern. Med.* **2025**, *12*, 005094. [CrossRef]
15. Annavajhala, M.K.; Gomez-Simmonds, A.; Uhlemann, A.-C. Multidrug-Resistant Enterobacter Cloacae Complex Emerging as a Global, Diversifying Threat. *Front. Microbiol.* **2019**, *10*, 44. [CrossRef]
16. Böhm, L.; Schaller, M.E.; Balczun, C.; Krüger, A.; Schummel, T.; Ammon, A.; Klein, N.; Helbing, D.L.; Eming, R.; Fuchs, F. A Case of Pseudomonas Straminea Blood Stream Infection in an Elderly Woman with Cellulitis. *Infect. Dis. Rep.* **2024**, *16*, 699–706. [CrossRef]
17. Tohya, M.; Tada, T.; Watanabe, S.; Kuwahara-Arai, K.; Zin, K.N.; Zaw, N.N.; Aung, M.Y.; Mya, S.; Zan, K.N.; Kirikae, T.; et al. Emergence of Carbapenem-Resistant Pseudomonas Asiatica Producing NDM-1 and VIM-2 Metallo-β-Lactamases in Myanmar. *Antimicrob. Agents Chemother.* **2019**, *63*, e00475-19. [CrossRef]
18. Liu, S.; Huang, N.; Zhou, C.; Lin, Y.; Zhang, Y.; Wang, L.; Zheng, X.; Zhou, T.; Wang, Z. Molecular Mechanisms and Epidemiology of Carbapenem-Resistant Enterobacter Cloacae Complex Isolated from Chinese Patients During 2004–2018. *Infect. Drug Resist.* **2021**, *14*, 3647–3658. [CrossRef]
19. Wenger, P.N.; Tokars, J.I.; Brennan, P.; Samel, C.; Bland, L.; Miller, M.; Carson, L.; Arduino, M.; Edelstein, P.; Aguero, S.; et al. An Outbreak of *Enterobacter hormaechei* Infection and Colonization in an Intensive Care Nursery. *Clin. Infect. Dis.* **1997**, *24*, 1243–1244. [CrossRef]
20. Tohya, M.; Watanabe, S.; Teramoto, K.; Uechi, K.; Tada, T.; Kuwahara-Arai, K.; Kinjo, T.; Maeda, S.; Nakasone, I.; Zaw, N.N.; et al. *Pseudomonas asiatica* sp. Nov., Isolated from Hospitalized Patients in Japan and Myanmar. *Int. J. Syst. Evol. Microbiol.* **2019**, *69*, 1361–1368. [CrossRef]
21. Bao, D.; Huang, L.; Yan, J.; Li, Y.; Ruan, Z.; Jiang, T. First Identification of a Multidrug-Resistant Pseudomonas Putida Co-Carrying Five β-Lactam Resistance Genes Recovered from a Urinary Tract Infection in China. *Infect. Drug Resist.* **2022**, *15*, 2229–2234. [CrossRef] [PubMed]
22. Brys, M.; Claerebout, E.; Chiers, K. Chronic Progressive Lymphedema in Belgian Draft Horses: Understanding and Managing a Challenging Disease. *Vet. Sci.* **2023**, *10*, 347. [CrossRef] [PubMed]
23. Fulop, T., Jr.; Jacob, M.P.; Robert, L. Determination of anti-elastin peptide antibodies in normal and arteriosclerotic human sera by ELISA. *J. Clin. Lab. Immunol.* **1989**, *30*, 69–74.
24. De Cock, H.E.; Van Brantegem, L.; Affolter, V.K.; Oosterlinck, M.; Ferraro, G.L.; Ducatelle, R. Quantitative and qualitative evaluation of dermal elastin of draught horses with chronic progressive lymphoedema. *J. Comp. Pathol.* **2009**, *140*, 132–139. [CrossRef] [PubMed]
25. Petersen, E.; Wågberg, F.; Ängquist, K.-A. Serum Concentrations of Elastin-Derived Peptides in Patients with Specific Manifestations of Atherosclerotic Disease. *Eur. J. Vasc. Endovasc. Surg.* **2002**, *24*, 440–444. [CrossRef]

26. Van Brantegem, L.; De Cock, H.E.V.; Affolter, V.K.; Duchateau, L.; Hoogewijs, M.K.; Govaere, J.; Ferraro, G.L.; Ducatelle, R. Antibodies to Elastin Peptides in Sera of Belgian Draught Horses with Chronic Progressive Lymphoedema. *Equine Vet. J.* **2007**, *39*, 418–421. [CrossRef]
27. De Keyser, K.; Berth, M.; Christensen, N.; Willaert, S.; Janssens, S.; Ducatelle, R.; Goddeeris, B.M.; De Cock, H.E.; Buys, N. Assessment of plasma anti-elastin antibodies for use as a diagnostic aid for chronic progressive lymphoedema in Belgian Draught Horses. *Vet. Immunol. Immunopathol.* **2015**, *163*, 16–22. [CrossRef]
28. Powell, H.; Affolter, V.K. Combined Decongestive Therapy Including Equine Manual Lymph Drainage to Assist Management of Chronic Progressive Lymphoedema in Draught Horses. *Equine Vet. Educ.* **2012**, *24*, 81–89. [CrossRef]

veterinary sciences

MDPI

Article

Circulating ACTH and Cortisol Investigations in Standardbred Racehorses Under Training and Racing Sessions

Cristina Cravana [1], Pietro Medica [1,*], Esterina Fazio [1], Katiuska Satué [2], Giacoma Brancato [1], Deborah La Fauci [1] and Giuseppe Bruschetta [1]

[1] Unit of Veterinary Physiology, Department of Veterinary Sciences, Messina University, Polo Universitario Annunziata, 98168 Messina, Italy; ccravana@unime.it (C.C.); fazio@unime.it (E.F.); giacoma.brancato@studenti.unime.it (G.B.); deborah.lafauci@unime.it (D.L.F.)

[2] Department of Animal Medicine and Surgery, Faculty of Veterinary, Cardenal Herrera-CEU University, Alfara del Patriarca, 46115 Valencia, Spain; ksatue@uchceu.es

* Correspondence: pmedica@unime.it; Tel.: +39-90-6766943

Simple Summary: This is an experimental study carried out on 10 trained Standardbreds, aged two and three years, including three females and seven males, with two main objectives: firstly, to examine the adrenocorticotropin (ACTH) and cortisol responses to training and racing sessions at rest condition, and at 5 min and 30 min after the training and racing sessions; secondly, to evaluate the effect of age and sex on endocrine parameters in both sessions. The effect of training and racing on ACTH ($p < 0.01$) and cortisol ($p < 0.01$) concentrations was obtained. Compared to the training session, horses showed greater ACTH concentrations at rest ($p < 0.001$), at 5 ($p < 0.01$) and 30 min ($p < 0.001$), and lower cortisol concentrations only at rest ($p < 0.01$) after racing; 2- and 3-year-old horses showed the greater ACTH concentrations at 5 and 30 min ($p < 0.01$) post-racing; males showed the greater ACTH concentrations at 5 min and 30 min ($p < 0.01$) post-racing.

Abstract: The hypothalamic-pituitary-adrenal (HPA) axis is a neuroendocrine system involved in the coping response to stressful challenges during exercise stimuli. Exercise represents a significant disruptor of homeostasis, inducing an ACTH-cortisol co-secretion, based on different characteristics of exercise in sport horses. Based on this statement, the aim of this study is to evaluate the circulating adrenocorticotropin and cortisol changes in Standardbred trotters, after training and racing sessions, considering the different age and sex. In particular, the aim is to determine to what extent the level of ACTH and cortisol increases during maximum effort in competition conditions (racing), and to compare two exercise conditions of different intensity, training and racing sessions, and effects on ACTH and cortisol responses. Ten Standardbreds, three females and seven males, clinically healthy, were enrolled and subjected to two exercise conditions: a non-competitive session (training) and then a competitive event (racing). Four of them were 2-year-olds and a further six were 3-year-olds. Training and racing effects on both ACTH ($p < 0.01$) and cortisol ($p < 0.01$) values were obtained. Compared to the training session, horses showed greater ACTH concentrations at rest ($p < 0.001$), at 5 ($p < 0.01$) and 30 min ($p < 0.001$), and lower cortisol concentrations only at rest ($p < 0.01$) after racing; 2- and 3-year-old horses showed the greater ACTH concentrations at 5 and 30 min ($p < 0.01$) post-racing; males showed the greater ACTH concentrations at 5 min and 30 min ($p < 0.01$) post-racing. The different stimuli of the two contexts, and differences in exercise intensity, such as training and competitive event, may have affected the direction of hypothalamic-pituitary-adrenal (HPA) axis response, both as an ability to adapt to physical stress of different intensity and as a preparatory activity for coping with stimuli. In conclusion, training and racing events induced a different HPA axis response in which both emotional experience and physical maturity could induce a significant adaptive response. As ACTH and cortisol concentrations in adult equids are extremely heterogeneous, further investigation is required to explore how different variables can influence the hormonal dynamics and their role as expressions of adaptive strategies to stress in horses.

Keywords: Standardbred; ACTH; cortisol; training; racing

1. Introduction

Exercise is a stressor that involves many regulatory endocrine systems that prompt the body to adapt and achieve a new dynamic equilibrium to maintain body homeostasis [1]. These adaptive endocrine responses are mainly expressed through the activation of the hypothalamic-pituitary-adrenal (HPA) axis contributing to include mechanical, metabolic, cardiovascular, and behavior modifications [2]. Several reports, conducted on sport horses and other athletic species in different exercise conditions, have highlighted the physiological and behavioral mechanisms by which HPA axis function influences stress adaptation. Primarily cortisol, among the HPA axis's hormones, increases hepatic gluconeogenesis and promotes lipolysis to provide fuel for prolonged, and submaximal exercise. Hence, ACTH and cortisol are both expressions of anabolic and catabolic balance and their concentrations often increase in response to stress and/or exhaustion; in fact, these conditions cause an increase in both circulating hormones [3]. Physical exercise represents a significant and powerful disruptor of homeostasis, inducing a consensual ACTH-cortisol co-secretion, based on the influence of type, intensity and duration of physical activity in sport horses [4–7], dogs and humans [8–10]. The individual fitness, the training degree [11], and age [12], and the novelty stimulus surge of the ACTH sets [13], along with the degree of previous experience in competitive racing [14] could affect the magnitude of ACTH and cortisol profiles. Moreover, the evaluation of HPA axis hormones has been used to assess training status, performance degree [15–20], and overtraining syndrome of athletic horses [12,18–28], sport humans [9,29], and, specifically, Standardbreds' performance [30–32]. Several reports have highlighted that long-term exercise increases cortisol concentrations, but short-term low-intensity exercise has not been shown in cortisol changes, or only slightly [33–35]. Interestingly, in the last decade the attention of scholars has been focused on the evaluation of salivary cortisol concentrations during different types of exercise and race training sessions [36–38]. Several reports have been published on how to reduce stress, and on the measurement of cortisol concentration through the non-invasive method of saliva [36,37,39–42]. Specifically, Strzleec et al. [36] reported that cortisol concentration decreases after dressage and that exercise intensity does not affect horses' cortisol concentrations, concluding that stress can be reduced by moderate intensity and duration of exercise. Therefore, differences in exercise intensity, such as those between walk, trot and light canter, do not influence the horse's stress. However, more research is needed to better understand the mechanisms of the adaptive responses of different types of sport horses during their athletic performance. On this basis, it has been hypothesized that different hormonal responses to non-competitive training and competitive sessions could occur, and that differences in exercise intensity may have affected the direction of hypothalamic-pituitary-adrenal (HPA) axis response, according to age and sex variables. Hence, the aim of this study was to determine to what extent the level of ACTH and cortisol in 2- and 3-year-old male and female Standardbreds increases during a race and to compare circulating ACTH and cortisol profiles after training and racing sessions.

2. Materials and Methods

2.1. Animals

All methods and procedures used in this study followed the guidelines of Italian law (D.L. 04/3/2014 n. 26) and the EU directive (2010/63/EU) on the protection of animals used for scientific purposes and approved by the Animal Ethics Committee for the Care and Use of Animals of University of Messina, Italy. (No. ME08/2023).

Two exercise conditions of different intensity (training and racing) were performed on 10 enrolled Standardbred racehorses: three females and seven males. Four of them were 2-years old and a further six were 3-years old and weighed (mean ± standard deviation) 384 ± 42 (range 331–485) Kg. All horses were kept in individual stalls bedded on wood shavings and allowed to move freely in a sand paddock 5 h/day. The horses were offered a diet consisting of grass (~8–10 kg/horse) and a commercial pelleted grain (~2 kg/horse) was fed and split into two feeds offered at 08.00 and 14.30 h. Horses were fed with diet formulated to meet the 1989 NRC for requirements horses. The concentrate portion of the ration contained 3.08 Mcal/kg digestible energy and 18% crude protein, while the hay contained 2 Mcal/kg digestible energy and 7% crude protein. Height at withers and total body weight were measured. For each horse, a body condition score (BCS) was evaluated on a scale of 1 to 9, with 1 being extremely emaciated and 9 extremely fat as reported by Henneke et al. [43]. Horses maintained body mass within 2–3% for a minimum of two weeks prior to the start of the study and all had body condition scores of 5–7 [43]. Horses remained healthy throughout the study with no change in BCS. Salt blocks and water were available ad libitum.

The general health state of the horses, based on thorough clinical and orthopedic examination, was assessed prior to the exercise 1 and 2. Physical examinations did not reveal clinical signs of any disease. All horses were in regular training. Although the participating horse and rider teams had a range of previous experience, all had competed in national rides in the past. This study was carried out during the spring, at the racecourse "La Favorita", located in Palermo (Sicily, Italy): 38°07′55″ N Longitude: 13°20′08″, elevation above sea level: 46 m, which has a 1000 m long oval racetrack. During the experiments, horses were exposed to similar environmental conditions; the external conditions were optimal with a dry and firm track, sunny but cool weather and no wind. Ambient conditions measured from times 14.00 h to 16.30 h at the racecourse were moderate throughout the course of both exercises. Air temperature ranged from 17 to 21.5 °C (the highest at being at 14.30 h); relative humidity ranged 40.5–55.5 (the highest being at 15.00 h). The weather conditions on the study days were similar, with a temperature of between 18 and 20 °C, moderate windiness (4–6 m/s) and relative humidity between 40 and 60%. Other data collected from the "ride cards" included heart rate (HR), respiratory rate (RR) and rectal temperature (RT).

Heart rate was measured using a pulsimeter (Polar S710i™, Polar Electro Oy, Kempele, Finland) continuously during both exercises. The HR at rest, at end of exercises and 15 and 30 min of recovery were recorded using a phonendoscope. Respiratory rate was recorded at rest, after the 500 m finish (within 2 min after finishing the exercise), and at 15 and 30 min post exercise.

The RT was measured by a commercial digital thermometer (Microlife Ag, Widnau, Switzerland) before both exercises at rest, at end of exercises, and at 15 and 30 min of recovery.

Horses were submitted to two different exercise conditions: a non-competitive training race (exercise 1) and then, after 3 days, a competitive event (exercise 2), according to similar protocols suitable for their performance activity. Both exercise sessions were performed between 14.00 and 17.30 h. The same drivers sat on a sulky towed by the horse both during training and racing sessions (Table 1).

Table 1. Scheme of training and racing protocols.

Training	Racing
Two rounds: strenuous training: velocity 8–10 m/s; duration 5 min	Two rounds: strenuous training: velocity 8–10 m/s; duration 5 min
1600 m: sprint training: velocity 10–12 m/s; duration 2.40 min	1600 m: sprint training: velocity 15–17 m/s; duration 1.58 min
Two rounds: basic training: velocity 5–8 m/s; duration 10 min	One round: basic training: velocity 5–8 m/s; duration 10 min
Cool down at the pass: duration 10 min	Cool down at the pass: duration 10 min

During the training session, the horses performed two racetrack rounds of "strenuous training" at the speed of 8–10 m/s for 5 min. Then, they took part in a "sprint training", during which they ran 1600 m, reaching a speed of 10–12 m/s. At the end, the horses carried out two "basic training" rounds at a speed of 5–8 m/s. For each phase of training, the driver kept track of speed to keep it constant using the distance marked by the stakes on the sidelines. At the end of the physical activity, each horse was brought to the stable, where the harness was removed. At this stage, each horse was submitted to a 10 min walking cool-down phase. Therefore, the full training lasted 30 min, including the cool-down phase.

The racing session was performed over 1600 m. Before the race, each horse performed a 5 min warm-up at the speed of 8–10 m/s, making two racetrack rounds of sprint training. During the race, the driver pushed the horse to the maximal exercise level, reaching the highest average speed of 15–17 m/s at the initial and the arrival phases. Subsequently, a basic training round at the speed of 5–8 m/s was carried out. At this stage, the horse was submitted to a 10 min walking cool-down phase. Therefore, the physical activity during the race lasted 30 min, including the cool-down phase.

2.2. Blood Sampling and Hormone Analyses

Blood samples (10 mL) were collected from the jugular prior to both exercise sessions, in basal condition, at 2 p.m. and 5 min after the training and race sessions, before the cool-down phase at 5 p.m. At the end, samples were collected 30 min after the exercise, including the cool-down phase, at 5.30 p.m. On the day of blood sampling, no restraint was necessary as the horses were already familiar with handling procedures. Informed consent from horse owners was provided.

Plasma ACTH was measured in unextracted plasma samples, and the concentrations were analyzed in duplicate using a commercially available radioimmunoassay kit (ELSA-ACTH, CIS-BioInternational, Gif-sur-Yvette, France) that have been validated for use on equine samples [44]. The hormone assay used has a range for ACTH detected of 0–440 pmol/L. The sensitivity of the assay ACTH was 0.44 pmol/L. The intra-assay and interassay coefficients of variation were 6.0% and 15.0%, respectively.

To analyze cortisol concentrations, blood samples were centrifuged at 3000× g for 15 min and the obtained serum samples stored at −20 °C until analyzed. Total serum cortisol concentrations were analyzed in duplicate using a commercial competitive enzyme assay (Enzyme Immunoassays, Roche Diagnostics GmbH, Mannheim, Germany) and an automated analyzer (BRIO, SEAC, Rome, Italy). During the first incubation, the cortisol sample competed with cortisol conjugated to horse radish peroxidase for the specific sites of the antiserum coated on the wells. Following incubation, all unbound material was removed by aspiration and washing. The enzyme activity bound to the solid phase is inversely proportional to cortisol concentration in calibrators and samples and is made evident by incubating the wells with a chromogen solution (tetramethylbenzidine) in substrate-buffer. Colorimetric reading was carried out using a spectrophotometer at 450, 405 nm wavelength (Sirio S, SEAC, Florence, Italy). The assay sensitivity was 5 ng/mL. The intra- and interassay CVs were 4% and 6.9%, respectively. All assays were performed according to the manufacturer's instructions. All samples were immediately processed in the adjacent laboratory at the racecourse and then harvested and stored in polystyrene tubes at −20 °C until analysis within one week.

2.3. Statistical Analysis

Data were presented as mean ± standard deviation (S.D.) of duplicate measurements in tables with the respective international units (UI) for the parameters studied. Normality was verified in all the data, using the Kolmogorov Smirnov test. In order to account for the different physiological variables, statistical analysis was performed by one-way analysis of variance (ANOVA) to evaluate age, sex and exercise effects on hormonal changes over basal time points in all sport horses.

A two-way repeated measures analysis of variance (Two-way RM ANOVA) was applied on the physiological changes, in order to test the effects of training and racing sessions together with their possible interaction. Post-hoc comparisons were performed using Tukey's test. The level of significance was set at $p < 0.05$. All calculations were performed using the PRISM package versions 10.3.0 (GraphPad Software Inc., San Diego, CA, USA). Also, percentage differences (Δ%) between after vs. at rest values and between racing session vs. training session values were also calculated.

3. Results

Means (mean ± S.D.) of ACTH and cortisol concentrations are presented in Tables 2–4 and in Figures 1–3.

Table 2. Plasma ACTH and serum cortisol concentrations of Standardbred horses before and after training session.

	ACTH (pmol/L)				
	at Rest	**5 min**	**Δ%**	**30 min**	**Δ%**
Total	2.50 ± 0.89	12.91 ± 3.43 [b]	+416	4.40 ± 8.66 [b]	+76
2-year-old	2.78 ± 0.95	12.13 ± 3.66 [a]	+336	5.10 ± 1.80	+83
3-year-old	2.31 ± 0.78	13.40 ± 4.06 [a]	+480	4.04 ± 1.83 [a]	+75
males	2.48 ± 0.89	10.28 ± 1.89 [b]	+314	4.51 ± 1.51 [a]	+82
females	2.11 ± 0.94	16.81 ± 2.21 [aB]	+697	4.19 ± 2.72	+98
	cortisol (nmol/L)				
Total	498 ± 136	632 ± 151 [b]	+27	629 ± 133 [b]	+26
2-year-old	555 ± 82	678 ± 59 [A]	+22	659 ± 76 [A]	+19
3-year-old	460 ± 158	601 ± 190 [a]	+31	608 ± 165 [a]	+32
males	475 ± 153	606 ± 175 [a]	+27	619 ± 156 [a]	+30
females	552 ± 79	692 ± 54	+25	651 ± 79	+18

Letters indicate differences vs. at rest: [a] ($p < 0.01$); [b] ($p < 0.001$); vs. 3-year-old [A] ($p < 0.01$); vs. males [B] ($p < 0.01$).

Table 3. Plasma ACTH and serum cortisol concentrations of Standardbred horses before and after racing session.

	ACTH (pmol/L)				
	at Rest	**Δ%**	**5 min**	**Δ%**	**30 min**
Total	6.13 ± 7.79 [C]	+203	18.61 ± 26.82 [Cb]	+23	7.52 ± 1.54 [CD]
2-year-old	6.43 ± 6.13	+243	22.05 ± 8.10 [Ca]	+29	8.29 ± 2.0 [C]
3-year-old	5.92 ± 8.70	+176	16.32 ± 2.74 [C]	+31	7.74 ± 3.28 [C]

Table 3. *Cont.*

	ACTH (pmol/L)				
	at Rest	**Δ%**	**5 min**	**Δ%**	**30 min**
males	6.30 ± 3.52	+172	17.15 ± 3.05 [bC]	+32	8.35 ± 3.25 [C]
females	5.34 ± 12.35	+68	22.29 ± 10.41 [b]	+44	7.67 ± 2.41 [b]
	cortisol (nmol/L)				
Total	383 ± 121 [C]	+64	629 ± 141 [b]	+47	565 ± 126 [a]
2-year-old	375 ± 118	+48	57 ± 215 [a]	+32	494 ± 165 [a]
3-year-old	389 ± 132	+74	678 ± 77	+57	612 ± 77
males	375 ± 99	+56	587 ± 143 [a]	+47	552 ± 123 [a]
females	405 ± 187	+79	725 ± 82 [a]	+46	593 ± 162 [a]

Letters indicate differences vs. at rest: [a] ($p < 0.01$); [b] ($p < 0.001$); vs. training [C] ($p < 0.01$); [D] ($p < 0.001$).

Table 4. Mean ± S.D. for heart rate (HR), respiratory rate (RR) and rectal temperature (RT) at rest, at the end of training and racing sessions, and at 15 min and 30 min of recovery.

	Training			
	At Rest	**End**	**15 min**	**30 min**
HR (beats/min)	38 ± 10	180 ± 20 [a]	80 ± 15 [a]	65 ± 5 [a]
RR (breaths/min)	18 ± 8	80 ± 5 [a]	75 ± 10 [a]	40 ± 10 [a]
RT °C	37.5 ± 0.3	38.6 ± 0.3 [a]	38.5 ± 0.2 [a]	37.8 ± 0.4
	racing			
HR (beats/min)	39 ± 11	189 ± 30 [a]	84 ± 10 [a]	55 ± 5 [a]
RR (breaths/min)	18 ± 3	80 ± 5 [a]	78 ± 10 [a]	45 ± 8 [a]
RT °C	37.6 ± 0.3	39.2 ± 0.4 [a]	38.6 ± 0.3 [a]	37.8 ± 0.4

[a] Indicates a significant difference ($p < 0.05$) from resting values.

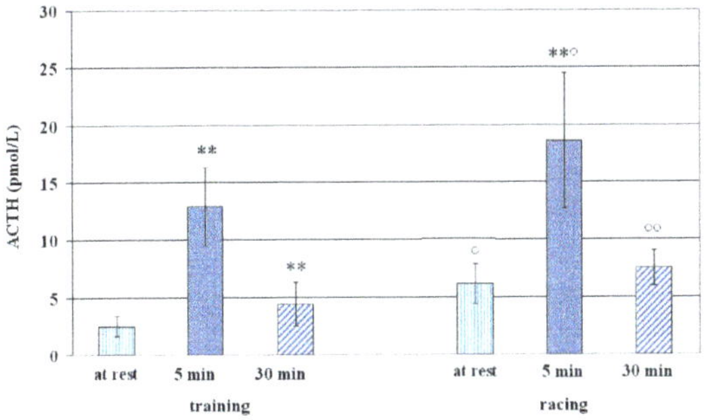

(a)

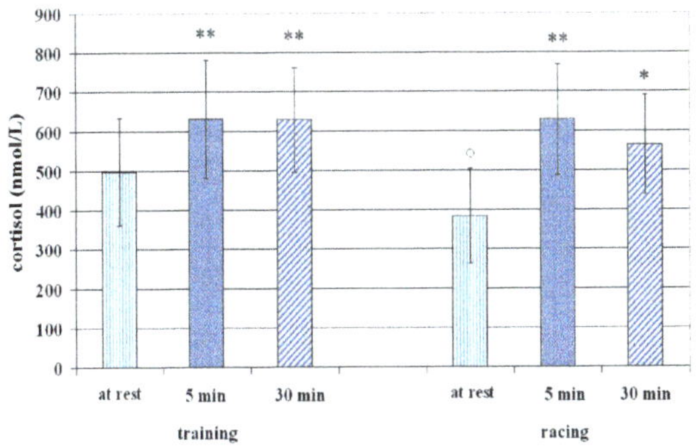

(b)

Figure 1. Circulating ACTH (**a**) and cortisol (**b**) concentrations (Mean ± S.D.) of Standardbreds before and after training and racing sessions. Asterisk indicates significant differences vs. at rest values * $p < 0.01$; ** $p < 0.001$. Symbol indicates significant differences vs. training ° $p < 0.01$; °° $p < 0.001$.

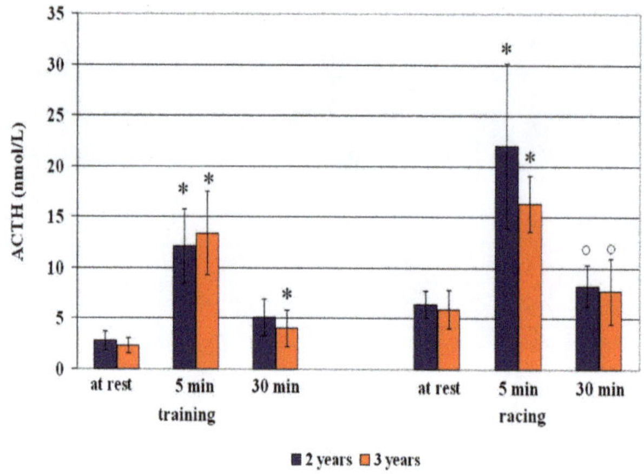

(a)

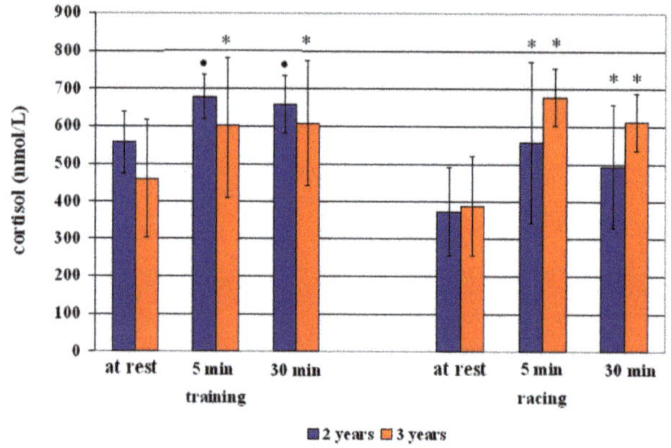

(b)

Figure 2. Circulating ACTH (**a**) and cortisol (**b**) concentrations (Mean ± S.D.) of Standardbred 2- and 3-year olds before and after training and racing sessions. Asterisk indicates significant differences vs. at rest values * $p < 0.01$. Symbol indicates significant differences vs. training ° $p < 0.01$. Symbol indicates significant differences vs. 3-years-old • $p < 0.01$.

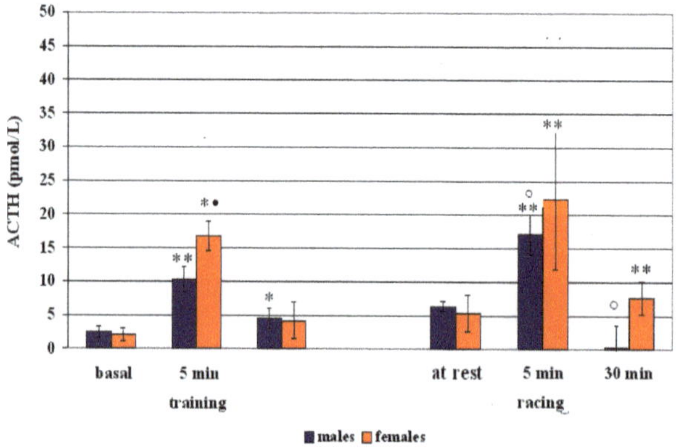

(a)

Figure 3. *Cont.*

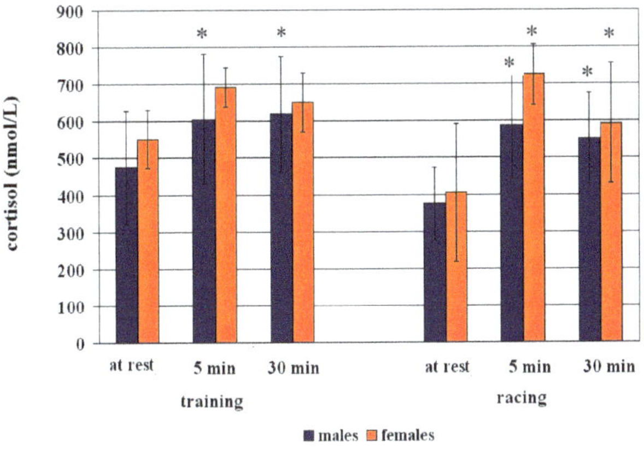

(b)

Figure 3. Circulating ACTH (**a**) and cortisol (**b**) concentrations (Mean ± S.D.) of male and female Standardbreds before and after training and racing sessions. Asterisk indicates significant differences vs. at rest values * $p < 0.01$, ** $p < 0.001$. Symbol indicates significant differences vs. racing 1 ° $p < 0.01$ Symbol indicates significant differences vs. males • $p < 0.01$.

3.1. ACTH-Cortisol Training Session Effect

The comparison of results obtained in this study with published data reported for horses in the literature did not reveal any large discrepancies and were in agreement with those physiological wide ranges; slight differences could be ascribed to differences in laboratory analysis techniques. However, data obtained confirm the trends recorded both in training and racing conditions [4–6,26–28,45–47].

Compared to resting values, a significant increase of ACTH concentrations 5 min (+414%; $p < 0.001$) and 30 min +76%; $p < 0.001$) after training, and of cortisol concentrations 5 min (+27%; $p < 0.001$) and 30 min (+66%; $p < 0.001$) after training was observed (Table 2 and Figure 1). A significant effect of training sessions both on the ACTH (F = 55.70; $p < 0.01$) and cortisol (F = 27.70; $p < 0.01$) changes was recorded.

Regarding the age effect, 2-year-old Standardbreds showed the greater ACTH concentrations at 5 min (+336%; $p < 0.01$) and at 30 min (+83%; $p < 0.01$), and, likewise, 3-year-olds showed greater concentrations at 5 min (+480%; $p < 0.01$) and 30 min (+75%; $p < 0.01$), compared to resting values. Significant effects of training on the ACTH concentrations both in 2-year-old horses (F = 28.73; $p < 0.01$) and 3-year-old ones (F = 29.80; $p < 0.01$) were obtained (Table 2 and Figure 1).

Three-year-old horses showed greater cortisol concentrations both at 5 min (+31%; $p < 0.01$) and 30 min (+32%; $p < 0.01$), compared to resting values. A significant effect of training on the cortisol concentrations only in 3-year-old horses (F = 39.57; $p < 0.01$) was obtained. Compared to 3-year-old horses, 2-year-old ones showed greater cortisol values ($p < 0.01$) at 5 min and 30 min.

Related to the sex effect, males showed greater ACTH concentrations at 5 min (+314%; $p < 0.001$) and 30 min (+82%; $p < 0.01$) after training, and females at 5 min (+697%; $p < 0.01$) and 30 min (+98%; $p < 0.01$) then resting values (Table 2 and Figure 1). A significant effect of training on the ACTH changes both in males (F = 47.30; $p < 0.01$) and in females (F = 10.66; $p < 0.01$) was obtained. The comparison between males and females showed the greatest ACTH concentrations at 5 min ($p < 0.01$) in females.

Related to cortisol concentrations, males showed greater concentrations both at 5 min (+27%; $p < 0.01$) and 30 min (+30%; $p < 0.01$), compared to resting values, with a significant effect of training on the cortisol of males (F = 19.02; $p < 0.01$) (Table 2 and Figure 1).

3.2. ACTH-Cortisol Racing Session Effect

Compared to resting values, an increase of ACTH concentrations at 5 min (+203%; $p < 0.001$) and 30 min (+23%; $p < 0.01$) after racing, and of cortisol at 5 min (+64%; $p < 0.001$) and 30 min (+47%; $p < 0.01$) after racing was observed (Table 3 and Figure 2). A significant effect on the ACTH ($F = 42.17$; $p < 0.01$) and cortisol ($F = 16.45$; $p < 0.01$) changes was recorded.

Regarding the age effect, both 2-year-old (+243%; $p < 0.01$) and older (+176%; $p < 0.001$) Standardbreds showed greater ACTH concentrations at 5 min ($p < 0.01$), compared to the resting values (Table 3 and Figure 2). A significant effect of racing both in 2-year-old horses ($F = 21.47$; $p < 0.01$) and in the older horses ($F = 31.41$; $p < 0.01$) was observed

Both 2-year-old and 3-year-old horses showed increases of cortisol concentrations at 5 min (+48% and +74%, respectively; $p < 0.01$) and 30 min (+32% and +57%, respectively; $p < 0.01$), compared to resting values (Table 3 and Figure 2). A significant racing effect on cortisol changes in younger ($F = 14.35$; $p < 0.01$) and older horses ($F = 7.75$; $p < 0.01$) was obtained.

Regarding the sex effect, males showed greater ACTH concentrations at 5 min (+172%; $p < 0.001$), and females both at 5 min (+68%; $p < 0.001$), and 30 min (+44%; $p < 0.001$), than resting values (Table 3 and Figure 2), with a significant effect of racing on the ACTH changes in males ($F = 49.95$; $p < 0.01$) and females ($F = 10.66$; $p < 0.01$).

Compared to resting values, males showed an increase in cortisol concentrations at 5 min +56%; ($p < 0.01$) and at 30 min (+47%; $p < 0.01$), and females showed an increase of cortisol concentrations at 5 min (+79%; $p < 0.01$) and at 30 min (+46%; $p < 0.01$) (Table 3 and Figure 2). A racing effect was observed on the cortisol changes in males ($F = 9.24$; $p < 0.001$) and females ($F = 27.87$; $p < 0.01$).

3.3. Racing Session vs. Training Session

Compared to the training session, horses showed greater ACTH concentrations at rest ($p < 0.001$), at 5 ($p < 0.01$) and 30 min ($p < 0.001$), and lower cortisol concentrations only at rest ($p < 0.01$) after racing; 2- and 3-year-old horses showed the greater ACTH concentrations at 5 and 30 min ($p < 0.01$) post-racing; males showed the greater ACTH concentrations at 5 min and 30 min ($p < 0.01$) post-racing (Table 3 and Figure 2).

3.4. Functional Variables

Heart rate (HR) and respiratory rate (RR) increased at the end of training and racing sessions, remaining greater at 15 min and 30 min compared to resting values ($p < 0.05$). Related to rectal temperature (RT), increases at the end and at 15 min were recorded returning to a t rest values at 30 min of recovery ($p < 0.05$) (Table 3).

4. Discussion

This research shows that different endocrine responses to non-competitive and competitive exercise conditions happen according to different ages and sex. ACTH and cortisol increase after the training and racing sessions, confirming previous data recorded in athletic horses after physical exercise, in which both hormones rise between 5 and 30 min after the end of exercise in Thoroughbred horses [23,48], Warmblood horses [16], and showjumping horses [4,25].

ACTH and cortisol concentrations, detected at rest time, were significantly greater, by +145% and +38%, respectively, prior to the racing than prior to the training session; this could be due to the psychological effect of waiting before the race start, as the role of affective processes underpinning temperament, mood and emotional reaction in determining discipline-specific performance [49].

The present study showed that ACTH concentrations at 5 min and 30 min post-racing exercise were 44% and 70% greater than at the same time post-training exercise. The same but more contained trend was recorded for cortisol concentrations at 5 min and 30 min post-exercise 2 that were 7% and 19% greater than at the same time post-training exercise.

A similar result was recorded by Nagata et al. [23], in which the greatest ACTH concentrations were observed at 5 and 30 min after the incremental treadmill exercise in Thoroughbred horses. In this study [23], plasma ACTH responses during exercise were more sensitive to the intensity of exercise than its duration.

Our results are in line with those reported in a recent study that showed an anticipatory response before racing in Standardbred racehorses, as shown by significant differences in pre-training (97.3 ± 16.4 nmol/L) and pre-race cortisol concentrations (171.8 ± 18.7 nmol/L), respectively [50]. Furthermore, other authors have shown that the participation of horses in equestrian competitions leads to an activation of the hypothalamus-adrenocortical function, reflected by a transient increase in cortisol release, and an increase in sympathoadrenal activity, indicated by a rise in heart rate [26].

In our experimental conditions, the training session speed was about 10–12 m/s and the racing session speed was about 15–17 m/s. Kurosawa et al. [22] reported that the onset time and the size of ACTH rise as a response to the SET on treadmill were significantly correlated to the speed sustained by Thoroughbred horses. Based on ACTH increase, Kurosawa et al. [22] stated the ACTH should be considered as a marker of stress and performance during physical exercise in Thoroughbred horses.

Our results are in agreement with Kurosawa et al. [22], since the HPA axis response of Standardbred was modulated by the increasing exercise intensity. In fact, our data showed a consistent effect of racing on the ACTH concentrations prior to the start, at 5 and 30 min after the race, compared to those of training. Moreover, previous studies found that not only ACTH response was more sensitive to the intensity of exercise, but it also showed a significant correlation with blood lactate concentrations [23] and with circulating arginine vasopressin concentrations during incremental exercise tests [13] Hada, 2003. The increase of ACTH values after 5 and 30 min of racing is also in agreement with ACTH increase recorded in Thoroughbreds submitted to the standardized exercise test (SET) on treadmill [16] Marc, 2000. However, submaximal treadmill exercises produced unmodified ACTH concentrations in young horses [20] McCarthy, 1991.

In young Friesian horses, a significant difference in the net increase of circulating ACTH concentrations was recorded after dressage training sessions; nevertheless, no difference in the post training values of cortisol concentrations was found [28] Fazio, 2016. Physical training clearly stimulated ACTH secretion, but, as previously recorded in a study by Ferlazzo et al. [25], an increase in circulating cortisol concentrations did not directly reflect differences in ACTH secretion, indicating that there was a differentiated response of the two hormones under the same conditions, probably related to different responses to mental exercise-related stress. The magnitude of ACTH changes after the first training session suggested that the novelty of stimuli represented an early stress manifestation in horses, thus inducing an initial, marked arousal response to stress, especially at the training start day; what is more, this result confirmed previous studies which showed that new stimuli, during exercise tests, increased plasma ACTH in Thoroughbred horses [13].

The participation of horses in competitive or non-competitive races caused an activation of HPA functions, reflected in a transient increase in cortisol release. Cortisol secretion is generally caused by an endogenous release following ACTH stimulation. This response during riding is at least in part due to the physical activity required of the horse. In agreement with previous studies, cortisol concentrations showed a significant increase after training and racing sessions. It is recognized that both a moderate exercise, like the training session of the present study, and an intense exercise, like the racing session, stimulate the cortisol releasing [50].

In fact, some authors showed a cortisol increase in response to an incremental exertion [19], after different types of sporting event competitions, including cross-country [50], endurance training sessions, competitions for Arabian horses [51], and after SET on the track [52]. Other authors have shown a 100% and 200% post-exercise increase in cortisol concentrations compared to rest values in horses submitted to endurance races [25] Ferlazzo, 2012. The increase of cortisol values at 5 and 30 min after racing is in agreement with a previously recorded cortisol rise after physical activity in trotters [53] Lindner 2002. Despite the different types of exercise, however, 5 min post-race the trend of cortisol, compared to that of ACTH, showed a rise comparable to that reported in post-training session. This result was also recorded by other authors that showed an increase of cortisol concentrations after dressage in both horses and riders [54].

The best recovery time was recorded during the training session; in fact, cortisol concentrations at 30 min post-racing exercise persisted at the high levels, quite comparable to those recorded at 5 min. This could be due to a major emotional impact and physical effort, characterizing a competitive race session. Previously, an increase of cortisol concentrations at 40–60 min post-race was reported in trotters [55]. Other authors showed that in Standardbred horses the cortisol significantly takes a cut 4–8 h after the exercise and then it needs about 24h to return at basal values [11]. Jumping horses showed a slower recovery of cortisol in the group of animals jumping the higher fences, with higher cortisol concentrations at 30 min post exercise than those of 5 min [25]. In addition, a significant cortisol increase before and after show jumping events in sport horses previously transported was recorded; moreover, the cortisol increases after the competitive event showed a correlation with the exercise rather than the transport stress [6]. Rivera et al. [56], reported high cortisol values at 15 min post-endurance training, with a decrease after 75 min.

Based on our results, the effect of circadian rhythms on ACTH and cortisol trends can be excluded; in fact, the blood samples were collected at the same times of the afternoon (from 3.00 to 5.30 p.m.). It is well known that exercise influences the regular circadian rhythms of cortisol secretion, which in the horse is greater in the morning than in the evening, with an acrophase at 10:50 a.m. [57]. A similar position is accepted in humans as well, since the cortisol response to exercise is significantly modulated by time of day; consequently, neglecting the circadian cortisol kinetic may introduce errors into conclusions about the hormonal response to exercise [58]. Regarding the age effect on ACTH and cortisol changes, the lower concentrations of both hormones in 2-year-olds than older horses at 5 min after the training session, and for cortisol at 5 and 30 min after, are in agreement with previous studies carried out on Standardbreds after SET on the oval dirt track [52]. Interestingly, albeit not significant, higher cortisol concentrations at all the times post-race compared to post-training in the younger horses were also recorded. One explanation could be due to the superimposed effort of both training and race sessions from each other. It is possible that the younger subjects spent a greater amount of energy than the older ones, which probably had better restorative ability and exercise adaptation capacities, as result of long-time training. A recent study showed that in Arabian racing horses, cortisol concentration tended to gradually increase after the subsequent training sessions during the racing season, and the best performances were achieved by the animals with lower cortisol concentrations [7]. After all, in human runners, the adaptation capacity of HPA axis was also demonstrated; cortisol concentrations were lower in trained than in untrained subjects under incremental exercise. Furthermore, the untrained controls showed an elevated cortisol concentration at the end of each intensity level; hence, trained athletes only had a significant increase in cortisol concentration after an exhaustive exercise session [20].

Age and training state positively influenced the locomotion parameters in Thoroughbreds and related metabolic costs related to competitive physical activity [59]. In addition, based on available data in the literature, Thoroughbred racehorses showed higher cortisol values in 2-year-olds than in the older horses [17]. It is possible to presume that the greater responsiveness of the HPA axis in the younger subjects was correlated to less experience in competitive race sessions, according to the greater emotional impact due to the novelty stress represented by the competitive event [13].

Although all horses in this work were adult, it is well documented that the age, genetic and environmental factors can impair the reactivity capacity of the HPA axis. The aging can also affect the hormonal responses usually put in place to cope with physical exercise, changing the control ability of the body on the cardiovascular function, the metabolism and substrate utilization [60]. Indeed, the authors showed that in Standardbred mares the HPA axis experience is submitted to a functional decline with age although the exact mechanisms remain unknown, while exercise training can facilitate the counteraction of these progressive deficits [30].

Regarding the influence of sex, the lack of significant differences in ACTH concentrations between males and females, during both training and racing, was reported. The cortisol concentrations showed the lower values in males compared to females, at all times and after both the training and the race sessions, but without a significant difference. The reason for this sex difference in the axis modulation in response to physical effort is not clear, so more investigations are needed.

Limitations of the Current Investigation

The clear limitation of this investigation is the small group of horses that completed the study, unfortunately related to events that occurred during the experiment. An additional 10 potential horses that were due to participate were excluded, due to retirement from racing, as were 10 because trainers did not want the horse to complete the exercises required by the study. Another limit is the lack of use of a heart rate monitor during both exercises to estimate speed and evaluate trot parameters, acceleration and distance covered, necessary for the assessment of fitness and performance. Furthermore, another limitation could be plasma cortisol sampling versus the non-invasive salivary cortisol method and the lack of some stress parameters (lactate, PCV, haematocrit). Unfortunately, this was a choice linked to the lack of authorization from race authorities and several owners and riders: it being a national competition forced us to reduce the analysis on the horses by allowing us only a single plasma sample from which it was possible to determine the measurement of both plasma parameters studied.

5. Conclusions

Different exercise intensities, in the two contexts of training and competition, induced the direction of the HPA axis response. This may influence the ability of horses to adapt to physical stress. It is reported in the literature that many factors can influence ACTH and cortisol levels. Measurement of basal ACTH levels is influenced by several pre-analytical factors and, in the context of low intensity exercise, significant increases have been reported after only 30 min of exercise. In order to optimize physical fitness in horses, further research is needed to explore how different variables (e.g., emotional experience, age, sex, sample management, etc.) may influence the ability to adopt stress coping strategies in performance horses.

Author Contributions: Conceptualization, C.C., E.F. and P.M.; formal analysis, C.C. and G.B. (Giacoma Brancato); investigation, C.C., K.S. and D.L.F.; data curation, G.B. (Giuseppe Bruschetta) and P.M.; writing—original draft preparation, C.C., P.M. and E.F.; writing—review and editing, E.F., C.C. and P.M.; visualization, C.C. and K.S.; supervision, K.S. and C.C. All authors have read and agreed to the published version of the manuscript.

Funding: This research received no external funding.

Institutional Review Board Statement: According to the mentioned legal regulations (study followed the guidelines of the Italian law (D.L. 04/3/2014 n. 26) and EU directive (2010/63/EU) on the protection of animals used for scientific purposes), the study cannot legally be classified as an animal experiment. In fact, the blood samplings are a common part of veterinary practice for the evaluation of sporting performance and anti-doping of horses in accordance with good veterinary practice, it does not need any separate ethical approval, according to Italian legislation.

Informed Consent Statement: Informed consent was obtained from all subjects involved in the study.

Data Availability Statement: The data that support this study will be shared upon reasonable request to be corresponding author.

Acknowledgments: We thank the trainers, drivers and all technical staff of the racecourse for their expert practical assistance. Furthermore, we would like to thank the veterinarian and all graduate and undergraduate students from University of Messina who assisted in collecting the data.

Conflicts of Interest: None of the authors of this paper has a financial or personal relationship with other people or organizations that could inappropriately influence or bias the content of the paper.

References

1. Chrousos, G.P. Stress and disorders of the stress system. *Nat. Rev. Endocrinol.* **2009**, *5*, 374–381. [CrossRef] [PubMed]
2. Mastorakos, G.; Pavlatou, M.; Diamanti-Kandarakis, E.; Chrousos, G.P. Exercise and the stress system. *Hormones* **2005**, *4*, 73–89. [PubMed]
3. Hill, E.E.; Zack, E.; Battaglini, C.; Viru, M.; Viru, A.; Hackney, A.C. Exercise and circulating cortisol levels: The intensity threshold effect. *J. Endocrinol. Investig.* **2008**, *31*, 587–591. [CrossRef]
4. Cravana, C.; Medica, P.; Prestopino, M.; Fazio, E.; Ferlazzo, A. Effects of competitive and noncompetitive show jumping on total and free iodothyronines, β-endorphin, ACTH and cortisol levels of horses. *Equine Vet. J.* **2010**, *42*, 179–184. [CrossRef]
5. Ferlazzo, A.; Cravana, C.; Fazio, E.; Medica, P. The different hormonal system during exercise stress coping in horses. *Vet. World* **2020**, *13*, 847–859. [CrossRef] [PubMed]
6. Ferlazzo, A.; Fazio, E.; Medica, P. Behavioral features and effects of transport procedures on endocrine variables of horses. *J. Vet. Behav.* **2020**, *39*, 21–31. [CrossRef]
7. Witkowska-Piłaszewicz, O.; Grzędzicka, J.; Sen, J.; Czopowicz, M.; Zmigrodzka, M.; Winnicka, A.; A Cywińska, A.; Carter, C. Stress response after race and endurance training sessions and competitions in Arabian horses. *Prev. Vet. Med.* **2021**, *188*, 105265. [CrossRef]
8. Angle, C.T.; Wakshlag, J.J.; Gillette, R.L.; Stokol, T.; Geske, S.; Adkins, T.O.; Gregor, C. Hematologic, serum biochemical, and cortisol changes associated with anticipation of exercise and short duration high-intensity exercise in sled dogs. *Vet. Clin. Pathol.* **2009**, *38*, 370–374. [CrossRef]
9. Duclos, M.; Tabarin, A. Exercise, training, and the Hypothalamo–Pituitary–Adrenal axis. *Front. Horm. Res.* **2016**, *47*, 12–26. [CrossRef]
10. Negro, S.; Bartolomé, E.; Molina, A.; Solé, M.; Gómez, M.D.; Valera, M. Stress level effects on sport performance during trotting races in Spanish trotter horses. *Res. Vet. Sci.* **2018**, *118*, 86–90. [CrossRef]
11. Golland, L.C.; Evans, D.L.; Stone, G.M.; Tyler-McGowan, C.M.; Hodgson, D.R.; Rose, R.J. Plasma cortisol and beta-endorphin concentrations in trained and over-trained Standardbred racehorses. *Pflug. Arch.* **1999**, *439*, 11–17. [CrossRef]
12. Malinowski, K.; Shock, E.J.; Rochelle, P.; Kearns, C.F.; Guirnalda, P.D.; McKeever, K.H. Plasma beta-endorphin, cortisol and immune responses to acute exercise are altered by age and exercise training in horses. *Equine Vet. J. Suppl.* **2006**, *36*, 267–273. [CrossRef] [PubMed]
13. Hada, T.; Onaka, T.; Takahashi, T.; Hiraga, A.; Yagi, K. Effects of novelty stress on neuroendocrine activities and running performance in thoroughbred horses. *J. Neuroendocrinol.* **2003**, *36*, 638–648. [CrossRef]

14. Cayado, P.; Muñoz-Escassi, B.; Domínguez, C.; Manley, W.; Olabarri, B.; Sánchez de la Muela, M.; Castejon, F.; Marañon, G.; Vara, E. Hormone response to training and competition in athletic horses. *Equine Vet. J. Suppl.* **2006**, *36*, 274–278. [CrossRef] [PubMed]
15. Thornton, J.R. Hormonal responses to exercise and training. *Vet. Clin. N. Am. Equine Pract.* **1985**, *1*, 477–496. [CrossRef]
16. Marc, M.; Parvizi, N.; Ellendorff, F.; Kallweit, E.; Elsaesser, F. Plasma cortisol and ACTH concentrations in the warmblood horse in response to a standardized treadmill exercise test as physiological markers for evaluation of training status in response to a standardized treadmill exercise test as physiological markers for evaluation of training status. *J. Anim. Sci.* **2000**, *78*, 1936–1946. [CrossRef]
17. Nogueira de Paula, G.; Barnabe, R.C.; Bedran-de-Castro, J.C.; Moreira, A.F.; Fernandes, W.R.; Mirandola, R.M.S.; How-ard, D.L. Serum cortisol, lactate and creatinine concentrations in Thoroughbreds fillies of different ages and states of training. *Braz. J. Vet. Res. Anim. Sci.* **2002**, *39*, 54–57. [CrossRef]
18. Quaranta, A.; Tateo, A.; Siniscalchi, M.; Padalino, B.; Iacoviello, R.; Centoducati, P. Influence of training on cortisol plasma levels and other hematic parameters in standardbred trotters. *Ippologia* **2006**, *17*, 5–10.
19. Peeters, M.; Joseph Sulon, J.; Didier Serteyn, D.; Vandenheede, M. Assessment of stress level in horses during competi-tion using salivary cortisol: Preliminary study. *J. Vet. Behav.* **2010**, *5*, 216. [CrossRef]
20. McCarthy, R.N.; Jeffcott, L.B.; Funder, J.W.; Fullerton, M.; Clarke, I.J. Plasma beta-endorphin and adrenocorticotrophin in young horses in training. *Aust. Vet. J.* **1991**, *68*, 359–361. [CrossRef]
21. Jimenez, M.; Hinchcliff, K.W.; Farris, J.W. Cathecolamine and cortisol responses of horses to incremental exertion. *Vet. Res. Commun.* **1998**, *22*, 107–118. [CrossRef]
22. Kurosawa, M.; Nagata, S.; Takeda, F.; Mima, K.; Hiraga, A.; Kai, M.; Taya, K. Plasma catecholamine, adrenocorticotropin and cortisol responses to exhaustive incremental treadmill exercise of the Thoroughbred horse. *J. Equine Sci.* **1998**, *9*, 9–18. [CrossRef]
23. Nagata, S.; Takeda, F.; Kurosawa, M.; Mima, K.; Hiraga, A.; Kai, M.; Taya, K. Plasma adrenocorticotropin, cortisol and catecholamines response to various exercises. *Equine Vet. J. Suppl.* **1999**, *30*, 570–574. [CrossRef]
24. de Graff-Roeflsema, E.; Keizer, H.A.; van Breda, E.; Wijnberg, I.D.; van der Kolk, J.H. Hormonal responses to acute exercise, training and overtraining. A review with emphasis on the horse. *Vet. Q.* **2007**, *29*, 82–101. [CrossRef] [PubMed]
25. Ferlazzo, A.; Medica, P.; Cravana, C.; Fazio, E. Circulating ß-endorphin, adrenocorticotropin, and cortisol concentrations of horses before and after competitive show jumping with different fence heights. *J. Equine Vet. Sci.* **2012**, *32*, 740–746. [CrossRef]
26. Becker-Bircka, M.; Schmidta, A.; Lasarzika, J.; Aurich, J.; Möstlc, E.; Aurich, C. Cortisol release and heart rate variability in sport horses participating in equestrian competitions. *J. Vet. Behav.* **2013**, *8*, 87–94. [CrossRef]
27. Bartolomé, E.; Cockram, M.S. Potential effects of stress on the performance of sport horses. *J. Equine Vet. Sci.* **2016**, *40*, 84–93. [CrossRef]
28. Fazio, E.; Medica, P.; Aveni, F.; Ferlazzo, A. The potential role of training sessions on the temporal and spatial physio-logical patterns in young Friesian horses. *J. Equine Vet. Sci.* **2016**, *47*, 84–89. [CrossRef]
29. Sokoloff, N.C.; Madhusmita, M.; Ackerman, K.E. Exercise, Training, and the Hypothalamic-Pituitary-Gonadal Axis in Men and Women. *Front. Horm. Res.* **2016**, *47*, 27–43. [CrossRef]
30. Liburt, N.R.; McKeever, K.H.; Malinowski, K.; Smarsh, D.N.; Geor, R.J. Response of the hypothalamic-pituitary-adrenal axis to stimulation tests before and after exercise training in old and young Standardbred mares. *J. Anim. Sci.* **2013**, *91*, 5208–5219. [CrossRef]
31. Bohák, Z.A.; Harnos, A.; Joó, K.; Szenci, O.; Kovács, L. Anticipatory response before competition in Standardbred racehorses. *PLoS ONE* **2018**, *13*, e0201691. [CrossRef]
32. Fazio, E.; Lindner, A.; Wegener, J.; Medica, P.; Hartmann, U.; Ferlazzo, A. Plasma cortisol concentration during standardized exercise in Standardbred racehorses within a racing season. *Pferdeheilkunde* **2023**, *39*, 151–157. [CrossRef]
33. Tsai, C.L.; Wang, C.H.; Pan, C.Y.; Chen, F.C.; Huang, T.H.; Chou, F.Y. Executive function and endocrinological responses to acute resistance exercise. *Front. Behav. Neurosci.* **2014**, *8*, 262. [CrossRef] [PubMed]
34. Wilk, M.; Krzysztofik, M.; Petr, M.; Zając, A.; Stastny, P. The slow exercise tempo elicits higher glycolytic and muscle dam age but not endocrine response that conventional squat. *Neuroendocrinol. Lett.* **2020**, *41*, 101–107.
35. Gordon, B.A.; Taylor, C.J.; Church, J.E.; Cousins, S.D. A comparison of the gluco-regulatory responses to high-intensity interval exercise and resistance exercise. *Int. J. Environ. Res. Public Health* **2021**, *18*, 287. [CrossRef] [PubMed]
36. Strzelec, K.; Kankofer, M.; Pietrzak, S. Cortisol concentration in the saliva of horses subjected to different kinds of exercise. *Acta Vet. Brno* **2011**, *80*, 101–105. [CrossRef]
37. Kędzierski, W.; Cywińska, A.; Strzelec, K.; Kowalik, S. Changes in salivary and plasma cortisol levels in Purebred Arabian horses during race training session. *Anim. Sci. J.* **2014**, *85*, 313–317. [CrossRef]
38. Kang, O.D.; Lee, W.S. Changes in Salivary Cortisol Concentration in Horses during Different Types of Exercise. *Asian-Australas. J. Anim. Sci.* **2016**, *29*, 747–752. [CrossRef]
39. Van Der Kolk, J.H.; Nachreiner, R.F.; Scott, H.C.; Refsal, K.R.; Zanella, K.J. Salivary and plasma concentra-tion of cortisol in normal horses and horses with Cushing's disease. *Equine Vet. J.* **2001**, *33*, 211–213. [CrossRef]

40. Mostl, E.; Palme, R. Hormones as indicators of stress. *Domest. Anim. Endocrinol.* **2002**, *23*, 67–74. [CrossRef]
41. Peeters, M.; Coline, C.; Becker, J.F.; Ledoux, D.; Vandenheede, M. Comparison between blood serum and salivary cortisol concentrations in horses using an adrenocorticotropic hormone challenge. *Equine Vet. J.* **2011**, *43*, 487–493. [CrossRef] [PubMed]
42. Kędzierski, W.; Strzelec, K.; Cywinska, A.; Kowalik, S. Salivary cortisol concentration in exercised thoroughbred horses. *J. Equine Vet. Sci.* **2013**, *33*, 1106–1109. [CrossRef]
43. Henneke, D.R.; Potter, G.D.; Kreider, J.L.; Yeates, B.F. Relationship between condition score, physical measurements and body fat percentage in mares. *Equine Vet. J.* **1983**, *15*, 371–372. [CrossRef]
44. Ferlazzo, A.; Fazio, E.; Aronica, V.; Di Majo, R.; Medica, P.; Grasso, L. Circulating concentrations of b-endorphin, ACTH and cortisol in horses after jumping over fences of different size. In Proceedings of the Conference on Equine Sports Medicine and Science, Cordoba, Spain, 24–26 April 1998; pp. 53–56.
45. Orth, D.N.; Holscher, M.A.; Wilson, M.G.; Nicholson, W.E.; Plue, R.E.; Mount, C.D. Equine Cushing's disease: Plasma immunoreactive proopiolipomelanocortin peptide and cortisol levels basally and in response to diagnostic tests. *Endocrinology* **1982**, *110*, 1430–1441. [CrossRef]
46. Hodson, N.P.; Wright, J.A.; Hunt, J. The sympatho-adrenal system and plasma levels of adrenocorticotropic hormone, cortisol and catecholamines in equine grass sickness. *Vet. Rec.* **1986**, *118*, 148–150. [CrossRef] [PubMed]
47. Eades, S.C.; Bounous, D.L. *Laboratory Profile of Equine Diseases*; Mosby Year Book: Saint Louis, MI, USA, 1997; pp. 15–20.
48. Cravana, C.; Fazio, E.; Satué, K.; Brancato, G.; Medica, P.; La Fauci, D. ACTH and cortisol dynamics in young Thoroughbred racehorses under competitive and non-competitive sessions. *J. Equine Vet. Sci.* **2024**. *submitted*.
49. McBride, S.; Mills, D.S. Psychological factors affecting equine performance. *BMC Vet. Res.* **2012**, *8*, 180. [CrossRef]
50. McKenna, B.; Lembert, M.; Evans, J.A. A study of β-endorphin and cortisol levels in the exercising horse. In Proceedings of the 12th Conference Association for Equine Sports Medicine, Fallbrook, CA, USA, 13–16 March 1993; Foreman, J., Ed.; Veterinary Practice Publishing Comp: Santa Barbara, CA, USA, 1993; pp. 39–43.
51. Desmecht, D.; Linden, A.; Amory, H.; Art, T.; Lekeux, P. Relationship of plasma lactate production to cortisol release following completion of different types of sporting events in horses. *Vet. Res. Commun.* **1996**, *20*, 371–379. [CrossRef]
52. Cravana, C.; Medica, P.; Fazio, E.; Di Giovanni, F.; Ferlazzo, A. Adrenocorticotropin and cortisol response to competitive and not competitive races in Thoroughbreds of different age. In *Management of Lameness Causes in Sport Horses: Muscle, Tendon, Joint and Bone Disorders*; Wageningen Academic Publishers: Wageningen, The Netherlands, 2006; pp. 165–168. [CrossRef]
53. Lindner, A.; Fazio, E.; Medica, P.; Ferlazzo, A. Effect of age, time record and V4 on plasma cortisol concentration in Standardbred racehorses during exercise. *Pferdeheilkunde* **2002**, *18*, 51–56. [CrossRef]
54. von Lewinski, M.; Biau, S.; Erber, R.; Ille, N.; Aurich, J.; Faure, J.M.; Möstl, E.; Aurich, C. Cortisol release, heart rate and heart rate variability in the horse and its rider: Different responses to training and performance. *Vet. J.* **2013**, *197*, 229–232. [CrossRef]
55. Nesse, L.; Johansen, G.I.; Blom, A.K. Effect of racing of lymphocyte proliferation in horses. *Am. J. Vet. Res.* **2002**, *63*, 528–530. [CrossRef] [PubMed]
56. Rivera, E.; Benjamin, S.; Nielsen, B.; Shelle, J.; Zanella, A.J. Behavioural and physiological responses of horses to initial training: The comparison between pastured versus stalled horses. *Appl. Anim. Behav. Sci.* **2002**, *78*, 235–252. [CrossRef]
57. Bohák, Z.; Szabó, F.; Beckers, J.-F.; Melo de Sousa, N.; Kutasi, O.; Nagy, K.; Szenci, O. Monitoring the circadian rhythm of serum and salivary cortisol concentrations in the horse. *Domest. Anim. Endocrinol.* **2013**, *45*, 38–42. [CrossRef] [PubMed]
58. Duclos, M.; Tabarin, A. *Activation of the HPA Axis During an Acute Bout of Exercise Chapter 2 Exercise, Training, and the Hypothalamo–Pituitary–Adrenal Axis*; Ghigo, E., Lanfranco, F., Strasburger, J., Eds.; Hormone Use and Abuse by Athletes, Endocrine Updates 29, 9; Springer Science and Business Media: Berlin/Heidelberg, Germany, 2011; pp. 9–11. [CrossRef]
59. Leleu, C.; Gloria, G.; Renault, G.; Barrey, E. Analysis of trotter gait on the track by accelerometry and image analysis. *Equine Vet. J. Suppl.* **2002**, *34*, 344–348. [CrossRef]
60. Traustadóttir, T.; Bosch, P.R.; Cantu, T.; Matt, K.S. Hypothalamic-pituitary-adrenal axis response and recovery from high-intensity exercise in women: Effects of aging and fitness. *J. Clin. Endocrinol. Metab.* **2004**, *89*, 3248–3254. [CrossRef]

veterinary sciences

MDPI

Communication

Characterization and Pathogenicity of Equine Herpesvirus Type 8 Using In-Vitro and In-Vivo Models

Yanfei Ji, Dandan Xu, Wenxuan Si, Yu Zhang, Muhammad Zahoor Khan, Xia Zhao * and Wenqiang Liu *

College of Agriculture and Biology, Liaocheng University, Liaocheng 252000, China
* Correspondence: zhaoxia@lcu.edu.cn (X.Z.); liuwenqiang@lcu.edu.cn (W.L.)

Simple Summary: Equine herpesvirus 8 (EHV-8) is one of the most important pathogens affecting donkeys, but there are no studies on the transmission potential, pathogenicity, and immune response of systemic EHV-8. We investigated the pathogenicity and immune response to major target organs by studying EHV-8 infection in a BALB/c mouse model. The results showed that EHV-8 was able to effectively replicate and cause pathological changes in the lungs, liver, and brain and elicited an immune response in the organism, providing basic data for the study of the pathogenesis of EHV-8.

Abstract: Equine herpesvirus type 8 (EHV-8) is predominantly isolated from donkeys, but its biological properties and pathogenic potential remain underexplored. This study aimed to investigate the biological characteristics and pathogenicity of the EHV-8 LCDC01 isolate by examining its effects in rabbit kidney (RK-13) cells and BALB/c mice. The virus was assessed for its ability to induce viral replication, pathological changes, and alterations in pro-inflammatory responses. In vitro, the EHV-8 infection of RK-13 cells induced characteristic cytopathic effects, including cell contraction, the formation of grapevine bundle-like structures, and detachment. In vivo, mice infected with the virus exhibited no clinical signs other than weight loss. Polymerase chain reaction (PCR) analysis detected viral DNA exclusively in the lungs of infected mice, while TaqMan PCR further confirmed the presence of EHV-8 nucleic acids in the lungs, liver, brain, and intestines. Furthermore, ELISA assays revealed a significant increase in the secretion of pro-inflammatory cytokines, including IL-1β, IL-6, IL-8, and IFN-α, in the lungs ($p < 0.05$). These findings suggest that EHV-8 primarily replicates in the lung tissue of mice and can induce inflammatory responses. This study provides valuable insights into the pathogenic mechanisms of EHV-8 and lays the groundwork for further investigation into its potential impact on equine and other animal populations.

Keywords: EHV-8; RK-13 cells cytopathic effects; pro-inflammatory cytokines; pathogenicity; BALB/c mice

1. Introduction

Equine herpesviruses (EHVs) comprise nine distinct subtypes, taxonomically classified into the *α-herpesvirus* and *γ-herpesvirus* subfamilies. Within this classification, EHV-1, EHV-3, EHV-4, EHV-8, and EHV-9 are members of the *α-herpesvirus* subfamily, with EHV-1, EHV-4, and EHV-8 representing the most clinically significant viruses in equine populations [1–3]. EHV-8, which predominantly affects donkeys, has been implicated in multiple clinical manifestations, including respiratory disease, reproductive failure with abortion [3], and neurological disorders [4]. Notably, EHV-8 demonstrates serological cross-reactivity with both EHV-1 and EHV-4, complicating diagnostic differentiation.

Epidemiological surveillance has revealed that EHV-8 is endemic in large-scale donkey breeding operations across Shandong Province, China, exhibiting distinct seasonal epidemic patterns that impose a substantial economic burden on the expanding donkey farming industry [5]. The rapid growth of donkey husbandry in China has elevated the importance of understanding EHV-8 pathobiology, rendering its study particularly relevant in comparison to EHV-1 and EHV-4 within this regional context.

Despite this increasing significance, research on EHV-8 remains comparatively limited, with the existing literature primarily focused on virus isolation, identification, and genomic characterization. Consequently, fundamental aspects of EHV-8 pathogenesis remain poorly elucidated, representing a critical knowledge gap. Previous investigations have utilized rabbit kidney epithelial (RK-13) cell culture systems followed by murine models to evaluate potential therapeutic interventions against EHV-8 infection [6]. Building upon this foundation, the present study examines the pathogenicity and inflammatory responses induced by EHV-8 in both RK-13 cells and a mouse model. The findings reported herein provide essential insights that may serve as a foundation for subsequent investigations into the molecular mechanisms governing EHV-8 infection and host-pathogen interactions.

2. Materials and Methods

2.1. Virus, Cells, and Mice

The EHV-8 strain LCDC01 (GenBank accession: PRJNA787358) was isolated in 2021 from nasal swabs collected from infected donkeys at a large-scale donkey farm in Liaocheng, China. Virus isolation was accomplished through the inoculation of RK-13 cells. The RK-13 cell lines were obtained from Wuhan Punosai Life Science and Technology Co., Ltd. (Wuhan, China). Cells were maintained in Dulbecco's Modified Eagle Medium (DMEM) supplemented with 10% fetal bovine serum (FBS). All cell culture reagents, including DMEM, FBS, and cell cryopreservation solution, were sourced from Dalian Meilun Biotechnology Co., Ltd. (Dalian, China). Specific pathogen-free (SPF) female BALB/c mice aged 2–3 weeks were used for the in vivo experiments. All mice were obtained from Jiangsu Huachuang Xinuo Pharmaceutical Technology Co., Ltd. (Taizhou, Jiangsu, China).

2.2. Viral Inoculation of Cells

RK-13 cells were cultured in 25 cm^2 flasks until approximately 80% confluence was reached. The virus was inoculated by adding 200 μL of EHV-8 virus suspension to the cells, followed by 800 μL of MEM medium (Meilun, Dalian, China). After a 1-h incubation at 37 °C, the viral inoculum was removed, and fresh complete medium (5 mL) was added. The cells were incubated overnight and periodically observed for cytopathic effects (CPE). Once CPE was apparent, the supernatant was collected to observe the viral morphology using a transmission electron microscope (H-7650, Hitachi Ltd., Tokyo, Japan).

2.3. Determination of $TCID_{50}$

RK-13 cells (1×10^7 Cells/mL) were seeded in 96-well plates and cultured at 37 °C in a 5% CO_2 atmosphere for 24 h. After reaching approximately 80% confluence, the culture medium was discarded. Virus inoculum derived from the original strain propagated in RK-13 cells (as described in Section 2.2) was subjected to serial 10-fold dilutions in MEM. Each dilution (100 μL) was added to wells in rows 1–11, while row 12 served as a negative control containing only MEM. Following incubation at 37 °C in a 5% CO_2 atmosphere for 1 h, the inoculum was removed and replaced with fresh complete medium. Cells were monitored daily for cytopathic effects (CPE). After 48 h post-infection (hpi), $TCID_{50}$ values were calculated based on the number of wells exhibiting CPE using the Reed–Muench method [7].

2.4. Immunofluorescence Detection of EHV-8 Antibody

An indirect immunofluorescence assay was performed according to the protocol described by Broeck et al. [8]. Following cell fixation, donkey anti-EHV-8 positive serum (diluted 1:100 in PBS) was applied (200 µL per well) and incubated for 2 h at room temperature. After three washes with PBS, FITC-conjugated rabbit anti-donkey secondary antibody (ImmunoWay, Plano, TX, USA) was added at a 1:1000 dilution and incubated for 1 h in darkness. Cells were subsequently mounted with 50 µL of 50% glycerol in blocking solution. The observation of fluorescence was made using a fluorescence microscope (Nikon Eclipse C1, Tokyo, Japan).

2.5. Apoptosis Detection

Apoptosis in EHV-8-infected RK-13 cells was assessed when approximately 30% of cells exhibited CPE using an Annexin V-FITC/ propidium iodide (PI) Apoptosis Detection Kit (Meron Biotechnology, Tianjin, China) according to the manufacturer's protocol. Briefly, adherent cells were harvested using EDTA-free trypsin, centrifuged at 1000× *g* for 5 min, and washed twice with cold PBS. Cell pellets were resuspended in 500 µL binding buffer, and 5 µL each of Annexin V-FITC and PI were added. Following gentle mixing and incubation for 15 min at room temperature in darkness, samples were analyzed by fluorescence microscopy.

2.6. Virus Infection and Sampling of Mice

BALB/c mice were randomly divided into two experimental groups: virus-infected (n = 15) and control (n = 15). Prior to inoculation, mice were anesthetized with CO_2 for 20 s. The virus-infected group received 50 µL of EHV-8 viral suspension ($TCID_{50}$ = $10^{-3.75}$/100 µL) intranasally, while the control group was administered 50 µL of sterile saline via the same route. Animals were monitored daily for clinical manifestations, and body weight was recorded at regular intervals. Blood samples were collected via retro-orbital puncture under appropriate anesthesia at 12 h, 1, 3, 5, and 7 days post-infection to evaluate the progression of EHV-8 infection. The serum was isolated by centrifugation of whole blood at 3000 × g for 10 min at 4 °C. At each time point, three mice from each group were euthanized by cervical dislocation. Pulmonary tissues were harvested for determination of the lung index, defined as the ratio of lung weight to body weight [9].

$$\text{Lung index} = (\text{Lung weight (g) Body weight (g)}) \times 100$$
$$\text{Lung index} = (\text{Body weight (g) Lung weight (g)}) \times 100$$

At 5 dpi, heart, liver, spleen, kidney, brain, and intestinal tissues were also collected for viral replication analysis to determine the damage of 5 days post-infection (dpi) to tissues and organs in mice. The lungs, liver, and brain were fixed with 4% formaldehyde and stained under the microscope according to the instructions of the HE (Hematoxylin-Eosin) Stain Kit (Solarbio, Beijing, China).

2.7. Determination of Viral Load in Major Organs of Mice

To determine the viral load at 5 dpi, we collected tissue samples from the heart, liver, spleen, lungs, kidneys, brain, and intestines from infected mice. Viral DNA (30 µL) was extracted from these 50 mg tissue samples using the AxyPrep Somatic Fluid Virus DNA/RNA Miniprep Kit (Corning Incorporated, New York, NY, USA). The extracted DNA was then analyzed by conventional PCR and quantitative PCR [10]. PCR primers for both conventional and quantitative PCR were designed based on the EHV-8 gB gene (GenBank accession: DVGE214585JN), with primer synthesis by Shanghai Bioengineering Co., Ltd. (Shanghai, China). The sequences of the primers are listed in Table 1.

Table 1. Primer Sequence Information.

Name	Sequences
PCR-Forward primer (F)	5′-TGTGAAAAATTCA AACGT-3′
PCR-Reverse primer (R)	5′-GAAGGTGCTGTTGCTTTTGCTGG-3′
Quantitative PCR-F	5′-GCGAACCCTCTGGAACGAAA-3′
Quantitative PCR-R	5′-TGGCTATCACGTCTCCCAGG-3′
Quantitative PCR-P	5′-ACTCTTGACGAGCGAGTTGCGGCGA-3′

The PCR reaction was performed in a total volume of 25 μL containing 12.5 μL of M5 Hiper Plus Taq HiFi DNA polymerase master mix (Mei5bio, Beijing, China), 1 μL of each primer (EHV-8-F and EHV-8-R, 10 μM each), 8.5 μL of nuclease-free water, and 2 μL of template DNA. Thermal cycling conditions were as follows: initial denaturation at 95 °C for 5 min; followed by 30 cycles of denaturation at 95 °C for 30 s, annealing at 60 °C for 30 s, then extend at 72 °C for 1 min; finally extend at 72 °C for 10 min.

For qPCR analysis, reactions were carried out in a 25 μL volume containing 12.5 μL of Premix Ex Taq (Takara, Japan), 1 μL of each primer (EHV-8-F and EHV-8-R, 10 μM each), 1 μL of EHV-8 probe (EHV-8-P, 10 μM), 7.5 μL of nuclease-free water, and 2 μL of template DNA. The qPCR thermal cycling parameters were: initial denaturation at 95 °C for 50 s, followed by 40 cycles of denaturation at 95 °C for 5 s and annealing/extension at 60 °C for 30 s. Cycle threshold (Ct) values were determined at the completion of the reaction and calculated based on the standard curve. Viral load quantification was achieved using a standard curve generated from serial dilutions of known concentrations of EHV-8 DNA.

2.8. Serological Detection of EHV-8 Antibodies

According to the EHV-8 gD antibody assay developed in our laboratory [11], mouse serum was applied to ELISA plates coated with polyclonal antibody to EHV-8 gD protein. The assay was performed in triplicate, and absorbance values were recorded at 450 nm.

2.9. Cytokine Detection

Mouse serum samples were analyzed for IL-1β, IL-6, IL-8, and IFN-α cytokine levels using Jang Su Enzyme-Linked Immunoassay kits and Enzyme-Linked Immunosorbent Assay (ELISA) kits following the manufacturer's protocols (Meimian, Yancheng, China). All assays were performed in technical triplicate. Absorbance values were measured at 450 nm (OD450) using a microplate reader to quantify cytokine concentrations.

2.10. Data and Analytics

Experimental data were counted and analyzed using SPSS statistics 23 (IBM, New York, NY, USA) and graphically analyzed using GraphPad prism 8.0.2 (GraphPad Software, Boston, MA, USA). Data are expressed as mean ± standard deviation (n = 3). Differences between the two groups were analyzed using the *t*-test, with "ns" indicating a non-significant difference $p > 0.05$, "*" indicating a significant difference $p < 0.05$, and "**" indicating a significant difference $p < 0.01$.

3. Results

3.1. Identification of the Virus

The EHV-8 strain was inoculated into monolayers of RK-13 cells and monitored daily via microscopy. After 48 h, the control group is normal (Figure 1a), and the infection group shows typical cytopathic effects, including cell rounding, fusion, and detachment (Figure 1b, orange arrow). Additionally, as shown in Figure 2, viral particles were examined using transmission electron microscopy, revealing particles approximately 100 nm in diameter, with a nearly circular shape and an outer vesicular envelope.

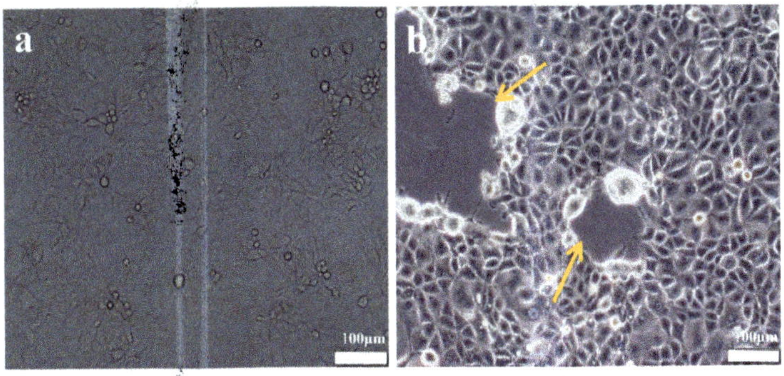

Figure 1. Cytopathological changes of EHV-8 infection (40×). (**a**): Control group, normal growth pattern of RK-13 cells in normal group. (**b**): EHV-8 infection group, orange arrow: EHV-8 infection of RK-13 cells with CPE phenomenon.

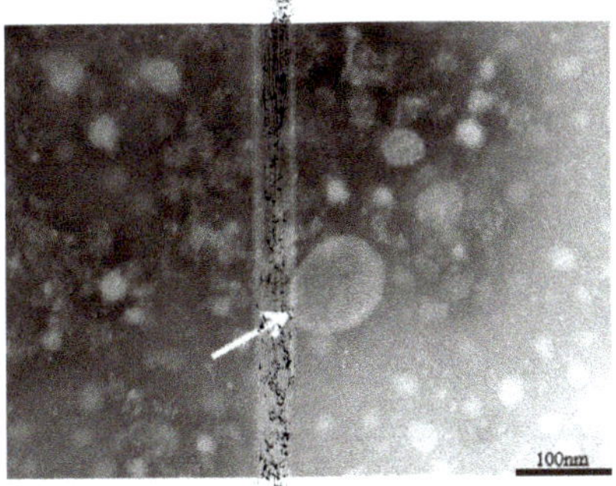

Figure 2. Electron microscopic observation of virus (×80.0 k). White arrow: EHV-8 virus particles.

3.2. *Viral Titer Measurement*

The viral stock was diluted at 1:10 and inoculated into RK-13 cells cultured in a 96-well plate. The CPE was monitored continuously, and wells exhibiting more than 50% CPE were counted. After 48 hpi, the determined $TCID_{50}$ of the isolated strains was $TCID_{50} = 10^{-3.75}/100\ \mu L$.

3.3. *Immunofluorescence Assay*

Figure 3 demonstrates EHV-8 infection and its effects on RK-13 cells. In Figure 3A, immunofluorescence analysis revealed abundant EHV-8-specific fluorescence in infected cells (Figure 3(Ab)), while control cells exhibited no detectable signal (Figure 3(Aa)), confirming efficient viral infection and replication in RK-13 cells. Figure 3B illustrates apoptotic events following viral infection. Early apoptosis, visualized by green fluorescence (white arrows), was markedly increased in infected cells (Figure 3(Bc)) compared to uninfected controls (Figure 3(Ba)). Similarly, late-stage apoptosis or necrosis, indicated by red fluorescence (yellow arrows), was substantially higher in infected cells (Figure 3(Bd)) than in control cells (Figure 3(Bb)). These results demonstrate that EHV-8 infection effectively induces apoptosis in RK-13 cells.

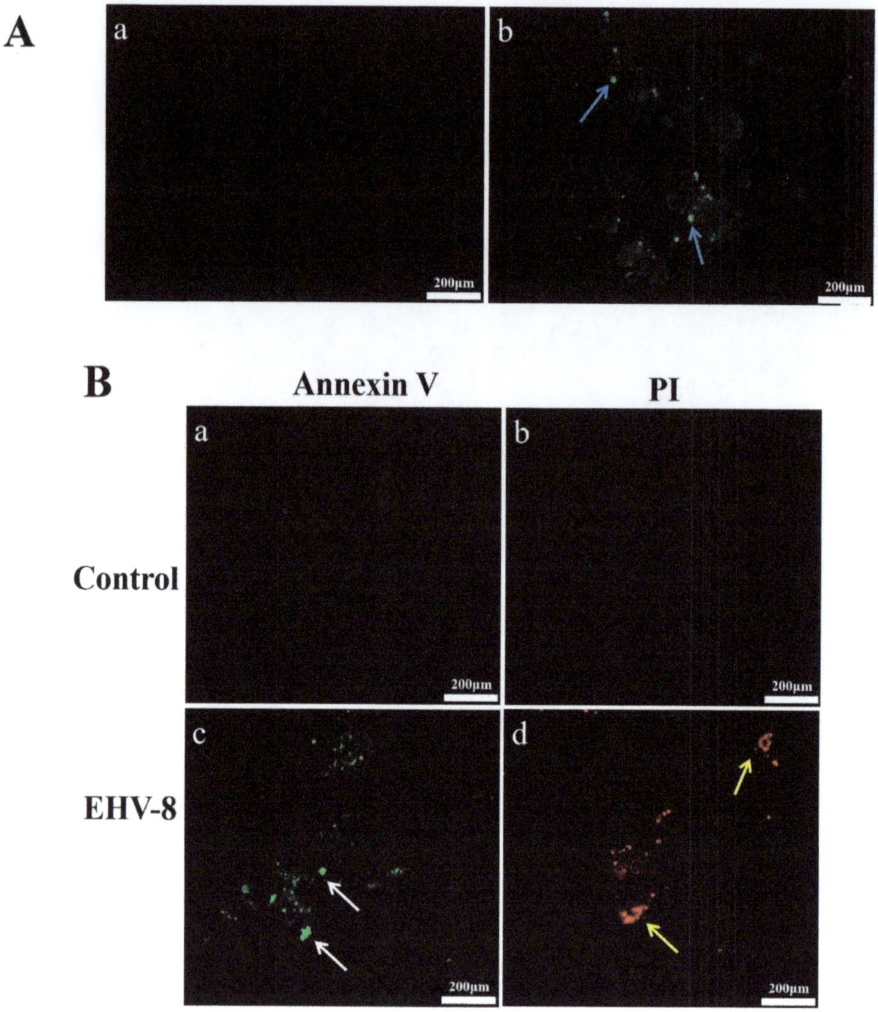

Figure 3. EHV-8 antibody test results and Apoptosis of EHV-8 infected cells and summary of immune responses (40×). (**Aa**) Indirect immunization with control EHV-8 antibody. (**Ab**) Indirect immunization with control EHV-8 antibody, blue arrows: EHV-8 binds to antibodies. (**Ba**) Control group: Annexin V fluorescence analysis of RK-13 cells. (**Bb**) Control group: Propidium iodide (PI) fluorescence analysis of RK-13 cells. (**Bc**) EHV-8 infection group: Annexin V fluorescence analysis of RK-13 cells, White arrows: early apoptotic cell fluorescence. (**Bd**) EHV-8 infection group: Propidium iodide (PI) fluorescence analysis of RK-13 cells yellow arrows: fluorescence of necrotic or advanced apoptotic cells.

3.4. Clinical and Pathological Changes in Mice

3.4.1. Effect of EHV-8 Infection on Body Weight in Mice

Following intranasal infection, no significant clinical symptoms were observed in the mice except for weight loss. Mice in the virus-infected group exhibited a gradual decrease in body weight from days 1 to 3 post-inoculation (dpi). From day 4 onwards, growth resumed, although at a slower rate compared to the control group. At 5 dpi, the growth rate of the infected mice returned to normal, while the control group continued to show normal weight gain (Figure 4).

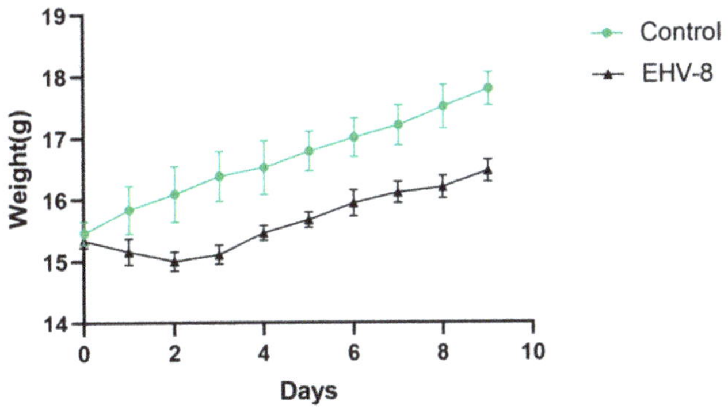

Figure 4. The line graph of body weight changes in mice. Note: Data are expressed as mean ± SEM (Standard Error of the Mean), and a *t*-test was used for differences between the two groups.

3.4.2. Effect of EHV-8 Infection on Lung Indices in Mice

As shown in Figure 5, there were no significant changes in the lung index at 12 h post- EHV-8 infection ($p > 0.05$). However, by day 5 post-infection, the lung index reached its peak at 2.48%, showing a highly significant difference compared to the control group ($p < 0.01$).

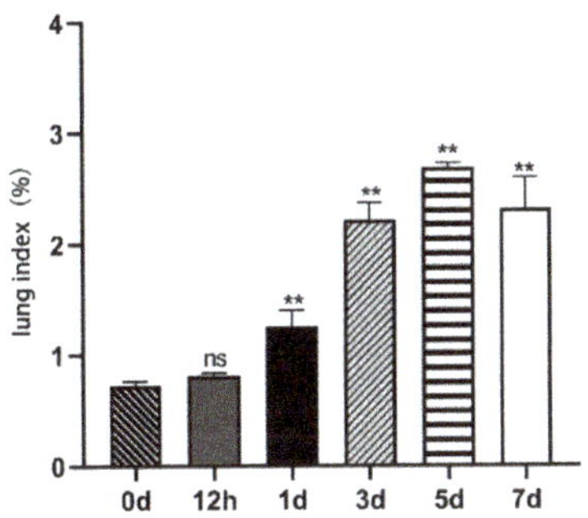

Figure 5. Lung indices at different times in EHV-8 infected mice. Note: Data are expressed as mean ± SEM (Standard Error of the Mean), and *t*-test was used for differences between the two groups, '**' indicates a significant extreme difference compared to the control group ($p < 0.01$); 'ns' indicates an insignificant difference compared to the control group ($p > 0.05$). (0 d): Lung index in control mice. (12 h): 12 h Lung index of mice infected for 12 h in EHV-8-infected group. (1 d): Lung index in mice infected with EHV-8 in the EHV-8-infected group for 1 day of infection. (3 d): Lung index of mice infected for 3 days in the EHV-8-infected group. (5 d): Lung index in mice infected with EHV-8 in the EHV-8-infected group for 5 days of infection. (7 d): Lung index of mice infected for 7 days in the EHV-8-infected group.

3.4.3. Pathological Observations of EHV-8 Infection on Mouse Organs

Pathological examination at 5 dpi revealed significant histopathological alterations in multiple organs of infected mice compared to controls. In the liver, control animals exhibited normal hepatic architecture (Figure 6b, green arrows), whereas infected mice demonstrated marked intravascular congestion, mild periportal steatosis, occasional cytoplasmic vacuolation in venous structures, and focal deposits of basophilic granular material. Notably, inflammatory cell infiltration was minimal in hepatic tissues. In the brain, control animals displayed normal neuronal morphology and organization (Figure 6d, red arrows). In contrast, infected mice exhibited numerous hippocampal neurons with hyperchromatic nuclei and disrupted cellular arrangement. Cytoplasmic vacuolization was prominent in neuronal cells, though glial proliferation was not observed. Pulmonary tissues from control animals maintained normal alveolar architecture, while infected mice demonstrated extensive alveolar septal thickening with significant inflammatory cell infiltration (Figure 6f, orange arrows).

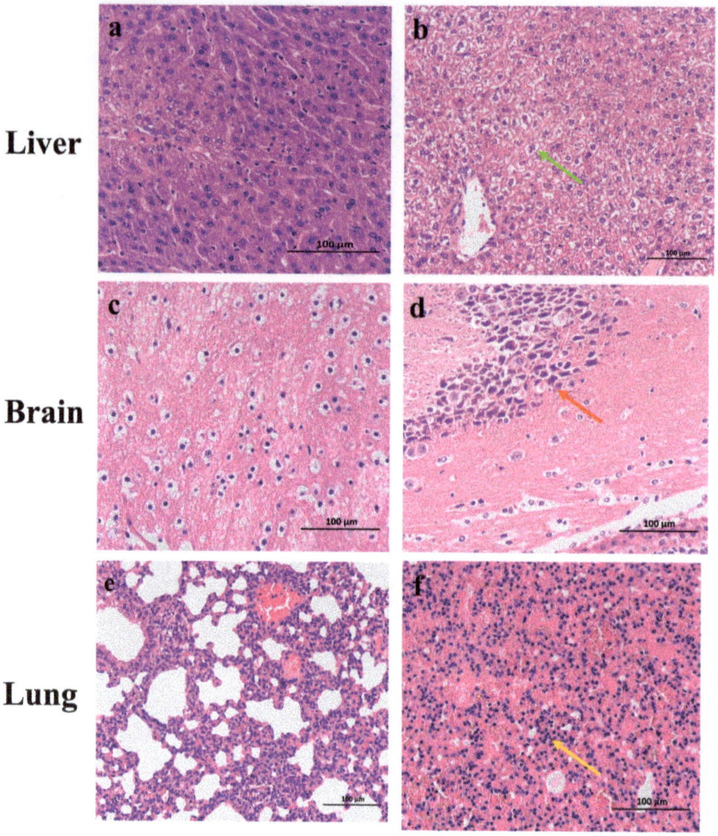

Figure 6. HE staining of various organs in the EHV-8 infected mice (40×). (**a**) HE staining of liver from mice in the control group. (**b**) HE staining of liver from mice in the EHV-8 infection group, green arrow: intravascular congestion, mild steatosis around the portal area, occasional vesicular vacuoles in the cytoplasm, and small clumps of bluish-purple granular material. (**c**) HE staining of brain from mice in the control group. (**d**) HE staining of brain from mice in the EHV-8 infection group, red arrow: vacuolization in the cytoplasm of nerve cells. (**e**) HE staining of lung from mice in the control group. (**f**) HE staining of lung from mice in the EHV-8 infection group, orange arrow: extensive thickening of the alveolar walls with numerous infiltrating inflammatory cells.

3.4.4. Distribution of EHV-8 Infected Mice in Different Organs and at Different Times of the Day

As demonstrated in Figure 7A, a 960 bp target band was detected exclusively in the lungs of mice at 5 dpi. No target bands were observed in other organs or control samples, confirming successful detection of EHV-8 nucleic acid specifically in the lungs of infected mice. Figure 7B illustrates the quantitative detection of EHV-8 viral load using fluorescence quantitative PCR. Viral DNA was detected in the liver, lung, brain, and intestine of 5 dpi mice, while remaining undetectable in all control samples. Notably, viral load in the lungs was significantly higher compared to the liver, brain, and intestine, suggesting preferential viral replication in pulmonary tissue.

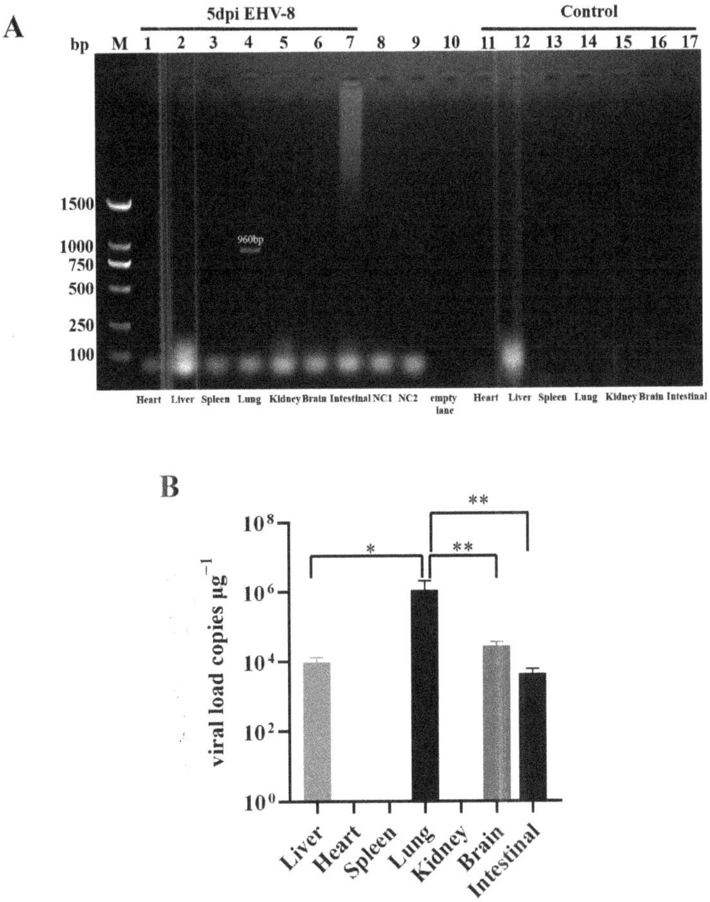

Figure 7. Levels of EHV-8 in mouse tissues after 5 d of infection. Note: Data are expressed as mean ± SEM (Standard Error of the Mean), and t-test was used for differences between the two groups, '*' indicates a significant difference compared to the control group ($p < 0.05$); '**' indicates a significant extreme difference compared to the control group ($p < 0.01$). (**A**): PCR result, samples are representative of each group, M: 2000 DNA maker; NC1: Negative control 1, NC2: Negative control 2. (**B**): 48 h EHV-8 group quantitative PCR result.

3.5. Temporal Dynamics of Anti-EHV-8 Immune Response

As shown in Figure 8, according to the EHV-8 gD antibody detection method, OD_{450} nm greater than 0.269 was judged as positive, after the mice were infected with EHV-8, the concentration of EHV-8 antibody increased gradually with the growth of time, and the concentration of antibody increased rapidly at 3–5 dpi, and there was a significant difference between the EHV-8 antibody concentration control group at 5 dpi, which indicated that EHV-8 was in the body of the mice, replicated, and induced an immune response in mice.

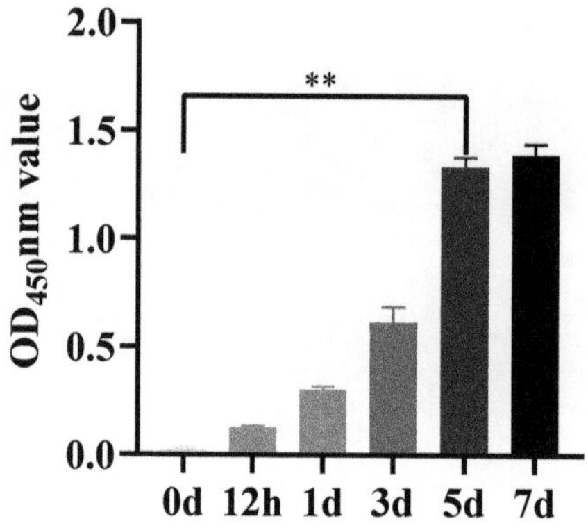

Figure 8. EHV-8 antibody level in mouse serum. Note: Data are expressed as mean ± SEM (Standard Error of the Mean), and t-test was used for differences between the two groups, '**' indicates a significant extreme difference compared to the control group ($p < 0.01$); (0 d): Control group. (12 h): Samples from 12 h of EHV-8 infection. (1 d): Samples from 1 day of EHV-8 infection. (3 d): Samples from 3 days of EHV-8 infection. (5 d): Samples from 5 days of EHV-8 infection. (7 d): Samples from 7 days of EHV-8 infection.

3.6. Cytokine Expression Patterns Following EHV-8 Infection

As shown in Figure 9, the secretion of cytokines, including IL-1β, IL-6, IL-8, and IFN-α, was significantly elevated in the lungs of infected mice compared to the control group ($p < 0.05$). The cytokine levels increased markedly at 3 dpi, peaked at 5 dpi, and subsequently decreased by 7 dpi.

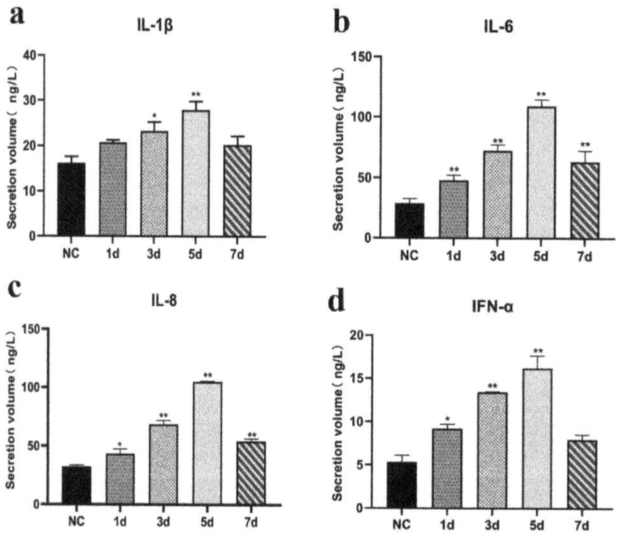

Figure 9. ELISA plot of immune factors for different days of EHV-8 infection. Note: Data are expressed as mean ± SEM (Standard Error of the Mean), and *t*-test was used for differences between the two groups, '*' indicates a significant difference compared to the control group ($p < 0.05$); '**' indicates a significant extreme difference compared to the control group ($p < 0.01$); (**a**): IL-1β ELISA result. (**b**): IL-6 ELISA result. (**c**): IL-8 ELISA result. (**d**): IFN-α ELISA result. (NC): negative control group. (1 d): Samples from 1 day of EHV-8 infection. (3 d): Samples from 3 days of EHV-8 infection. (5 d): Samples from 5 days of EHV-8 infection. (7 d): Samples from 7 days of EHV-8 infection.

4. Discussion

In the present investigation, we sought to comprehensively characterize viral proliferation, histopathological manifestations, and inflammatory responses in both RK-13 cells and a murine model following infection with the LCDC01 isolate. Upon infection, the LCDC01 strain consistently induced characteristic CPE and apoptotic changes in RK-13 cells, thereby providing compelling evidence of active viral replication within these cellular substrates. Notably, despite the absence of overt clinical manifestations in the infected mice, a significant reduction in body weight was documented, which strongly suggests that the LCDC01 strain possesses modest toxicity and demonstrates relatively low virulence in vivo.

The detailed histopathological examination of various tissue sections revealed differential patterns of inflammatory damage across organs, with pulmonary tissues exhibiting particularly pronounced pathological alterations compared to hepatic and cerebral tissues. These observations are in concordance with previously published findings regarding EHV-1-induced tissue pathology in murine models [12,13]. Furthermore, molecular analysis using PCR for the detection of EHV-8 nucleic acid yielded positive results exclusively in pulmonary specimens, whereas all other examined organs generated negative findings. This apparent discrepancy strongly indicates that the virus preferentially replicates within pulmonary tissues, with substantially reduced viral burdens in extrapulmonary sites. It is important to note that the inherent sensitivity limitations of conventional PCR methodology may preclude the detection of viral genomic material in tissues harboring low viral loads, a hypothesis supported by the positive results subsequently obtained using more sensitive quantitative PCR assays.

Following the experimental inoculation of RK-13 cells with the LCDC01 isolate, we observed the development of characteristic CPE at 48 hpi, with a calculated $TCID_{50}$ of $10^{-3.75}$ mL^{-1}. These values are consistent with previously reported metrics for EHV-1-induced cytopathology in comparable experimental systems [14]. Ultrastructural analysis using transmission electron microscopy (TEM) revealed the presence of typical herpesvirus virions measuring approximately 100 nm in diameter, which aligns precisely with the established morphological characteristics of EHV-like viruses described in the literature [15].

The complex interplay between viral infection and host immune response is critical in determining disease progression and severity. Thus, cytokines such as IL-1β, IL-6, IL-8, and IFN-α play fundamental roles in orchestrating the immune system's multifaceted response to injury, infection, and malignancy. Each of these signaling molecules contributes distinctively to immune regulation, tissue repair, and pathogenesis across various physiological contexts [16,17]. Consistent with previous research demonstrating that EHV-1 infection triggers upregulation of pro-inflammatory cytokines in murine models [18], our investigation revealed significant elevations in IL-1β, IL-6, IL-8, and IFN-α levels in both the serum and pulmonary tissues of mice infected with the LCDC01 strain. Importantly, the secretion of these inflammatory mediators showed marked increases, specifically on days 3 and 5 post-infection. This temporal pattern strongly suggests that the LCDC01 strain not only induces lung injury but also elicits a robust inflammatory cascade in the pulmonary tissues of infected mice, which is in line with previously reported research [6]. Collectively, these findings provide valuable insights into the pathogenic mechanisms underlying EHV-8 infection and highlight the pivotal role of inflammatory cytokines in viral pathogenesis. The temporal correlation between peak cytokine levels and histopathological damage underscores the potential contribution of immune-mediated mechanisms to tissue injury. Further investigations are warranted to elucidate the precise molecular pathways involved in EHV-8-mediated tissue damage and immune dysregulation, which may ultimately inform therapeutic strategies targeting specific components of the inflammatory response.

5. Conclusions

The EHV-8 represents a significant pathogen affecting donkeys with considerable economic implications for the donkey industry. In this study, we found that the LCDC01 strain EHV-8 was able to cause cytopathic effects and apoptosis in RK-13 cells. Upon infection in the murine model, EHV-8 caused substantial pathological damage to hepatic, neural, and pulmonary tissues, with concurrent elevation of serum immune mediators. Notably, pulmonary tissue exhibited the highest viral load among examined organs. These findings provide a foundational framework for understanding EHV-8 pathogenesis and host-pathogen interactions, which may inform future preventive and therapeutic strategies against this economically important equid herpesvirus.

Author Contributions: Conceptualization, Y.J., Y.Z., W.S., D.X., X.Z., and W.L.; Data curation, Y.J. and W.S.; Formal analysis, Y.J.; Investigation, Y.J., D.X., Y.Z., and W.L.; Methodology, Y.J., D.X., Y.Z., W.S., M.Z.K., X.Z., and W.L.; Project administration, X.Z. and W.L.; Resources, X.Z. and W.L.; Software, Y.J., D.X., Y.Z., W.S., and X.Z.; Supervision, X.Z. and W.L.; Validation, W.S., D.X., M.Z.K., and W.L.; Visualization, Y.J., D.X., M.Z.K., and W.L.; Writing—original draft, Y.J., D.X., X.Z., and W.L.; Writing—review & editing, Y.J., W.S., D.X., M.Z.K., X.Z., and W.L. All authors have read and agreed to the published version of the manuscript.

Funding: This research was funded by the Natural Science Foundation of Shandong Province (ZR2023MC209), Shandong Donkey Industry Technology System Fund (SDAIT-27-11), Liaocheng University Science and Technology Tackling Project (318011701, 318052242), Liaocheng Key R&D Program Policy Guidance Project (2022YDNY04).

Institutional Review Board Statement: All procedures were approved by the Animal Welfare and Ethics Committee of the Institute of Animal Science, Liaocheng University (Protocol No. 2023022712). The date of ethical approval was 27 February 2023.

Informed Consent Statement: Not applicable.

Data Availability Statement: The data supporting the conclusions of this case report are included in this article. All data sets can be requested from correspondence with the authors.

Acknowledgments: Many thanks to the subject group for their support and help with this experiment.

Conflicts of Interest: The authors declare that they have no known competing financial interests or personal relationships that could have appeared to influence the work reported in this paper.

References

1. El Brini, Z.; Cullinane, A.; Garvey, M.; Fassi Fihri, O.; Fellahi, S.; Amraoui, F.; Loutfi, C.; Sebbar, G.; Paillot, R.; Piro, M. First Molecular and Phylogenetic Characterization of Equine Herpesvirus-1 (EHV-1) and Equine Herpesvirus-4 (EHV-4) in Morocco. *Animals* **2025**, *15*, 102. [CrossRef] [PubMed]
2. Worku, A.; Molla, W.; Kenubih, A.; Negussie, H.; Admassu, B.; Ejo, M.; Dagnaw, G.G.; Bitew, A.B.; Fentahun, T.; Getnet, K.; et al. Molecular Detection of Equine Herpesviruses from Field Outbreaks in Donkeys in Northwest Amhara Region, Ethiopia. *Vet. Med. Int.* **2024**, *2024*, 9928835. [CrossRef] [PubMed]
3. Ruan, L.; Li, L.; Yang, R.; You, A.; Khan, M.Z.; Yu, Y.; Chen, L.; Li, Y.; Liu, G.; Wang, C.; et al. Equine Herpesvirus-1 Induced Respiratory Disease in Dezhou Donkey Foals: Case Study from China, 2024. *Vet. Sci.* **2025**, *12*, 56. [CrossRef] [PubMed]
4. Schvartz, G.; Edery, N.; Moss, L.; Hadad, R.; Steinman, A.; Karniely, S. Equid Herpesvirus 8 Isolated from an Adult Donkey in Israel. *J. Equine Vet. Sci.* **2020**, *94*, 103247. [CrossRef] [PubMed]
5. Chen, L.; Li, S.; Li, W.; Yu, Y.; Sun, Q.; Chen, W.; Zhou, H.; Wang, C.; Li, L.; Xu, M.; et al. Rutin prevents EqHV-8 induced infection and oxidative stress via Nrf2/HO-1 signaling pathway. *Front. Cell. Infect. Microbiol.* **2024**, *14*, 1386462. [CrossRef] [PubMed]
6. Hu, L.; Wang, T.; Ren, H.; Liu, W.; Li, Y.; Wang, C.; Li, L. Characterizing the Pathogenesis and Immune Response of Equine Herpesvirus 8 Infection in Lung of Mice. *Animals* **2022**, *12*, 2495. [CrossRef] [PubMed]
7. Mizuno, M.; Kimbara, S.; Ichise, H.; Ishikawa, N.; Nishihara, Y.; Nishio, M.; Sekiya, I. Cleaning methods for biosafety cabinet to eliminate residual mycoplasmas, viruses, and endotoxins after changeover. *Regen. Ther.* **2025**, *28*, 73–80. [CrossRef] [PubMed]

8. Broeckl, C.V.; Hiereth, S.; Straubinger, R.K. A comparative study evaluating three line immunoassays available for serodiagnosis of equine Lyme borreliosis: Detection of Borrelia burgdorferi sensu lato-specific antibodies in serum samples of vaccinated and non-vaccinated horses. *PLoS ONE* **2024**, *19*, e0316170. [CrossRef] [PubMed]
9. Wu, X.; Xu, L.; Xu, G.; Xu, Y.; Liu, H.; Hu, Y.; Ye, X.; Huang, Q.; Tang, C.; Duan, N.; et al. Fei-yan-qing-hua decoction exerts an anti-inflammatory role during influenza by inhibiting the infiltration of macrophages and neutrophils through NF-κB and p38 MAPK pathways. *J. Ethnopharmacol.* **2025**, *337*, 118846. [CrossRef] [PubMed]
10. Ji, Y.; Qi, L.; Zhang, J.; Xu, D.; Zhang, J.; Zhao, X.; Liu, W. Establishment of TaqMan Real-time Fluorescence Quantitative PCR Detection Method for Equine Herpesvirus Type 8. *Chin. Vet. J.* **2023**, *59*, 24–28.
11. Zhang, J. *Epidemiological Survey of Equine Herpesvirus in Large-Scale Donkey Farms and Preparation of gD Protein Polyantibodies of EHV-8 Isolates*; Liaocheng University: Liaocheng, China, 2022.
12. Kimura, H.; Shibata, M.; Kuzushima, K.; Nishikawa, K.; Nishiyama, Y.; Morishima, T. Detection and direct typing of herpes simplex virus by polymerase chain reaction. *Med. Microbiol. Immunol.* **1990**, *179*, 177–184. [CrossRef] [PubMed]
13. Walker, C.; Love, D.N.; Whalley, J.M. Comparison of the pathogenesis of acute equine herpesvirus 1 (EHV-1) infection in the horse and the mouse model: A review. *Vet. Microbiol.* **1999**, *68*, 3–13. [CrossRef] [PubMed]
14. Gosztonyi, G.; Borchers, K.; Ludwig, H. Pathogenesis of equine herpesvirus-1 infection in the mouse model. *Apmis* **2010**, *117*, 10–21. [CrossRef] [PubMed]
15. Tong, P.; Pan, J.; Dang, Y.; Yang, E.; Jia, C.; Duan, R.; Tian, S.; Palidan, N.; Kuang, L.; Wang, C.; et al. First identification and isolation of equine herpesvirus type 1 in aborted fetal lung tissues of donkeys. *Virol. J.* **2024**, *21*, 117. [CrossRef]
16. Heinrich, P.C.; Behrmann, I.; Haan, S.; Hermanns, H.M.; Müller-Newen, G.; Schaper, F. Principles of interleukin (IL)-6-type cytokine signalling and its regulation. *Biochem. J.* **2003**, *374*, 1–20. [CrossRef] [PubMed]
17. Krumm, B.; Xiang, Y.; Deng, J. Structural biology of the IL-1 superfamily: Key cytokines in the regulation of immune and inflammatory responses. *Protein Sci.* **2014**, *23*, 526–538. [CrossRef] [PubMed]
18. Shi, W.; Yao, X.; Fu, Y.; Wang, Y. Interferon-α and its effects on cancer cell apoptosis (Review). *Oncol. Lett.* **2022**, *24*, 235. [CrossRef]

veterinary sciences

MDPI

Article

Absolute Quantitative Lipidomics Reveals Differences in Lipid Compounds in the Blood of Trained and Untrained Yili Horses

Tongliang Wang [1,2,3], Jun Meng [1,2,3], Jianwen Wang [1,2,3], Wanlu Ren [1,2,3], Xixi Yang [1,2,3], Wusiman Adina [1], Yike Bao [1], Yaqi Zeng [1,2,3,*] and Xinkui Yao [1,2,3,*]

1 College of Animal Science, Xinjiang Agricultural University, Urumqi 830052, China; wtl13639911402@163.com (T.W.); junm86@xjau.edu.cn (J.M.); wjw1262022@126.com (J.W.); 13201295117@163.com (W.R.); xxyang2022@126.com (X.Y.); adina0901@126.com (W.A.); b18599635754@126.com (Y.B.)
2 Xinjiang Key Laboratory of Horse Breeding and Exercise Physiology, Urumqi 830052, China
3 Horse Industry Research Institute, Xinjiang Agricultural University, Urumqi 830052, China
* Correspondence: zengyaqi@xjau.edu.cn (Y.Z.); yxk61@126.com (X.Y.); Tel.: +86-189-9796-4076 (Y.Z.); +86-138-998-05781 (X.Y.)

Simple Summary: This study aims to understand how training affects the heart structure and blood fat metabolism of the Yili horse, known for its athletic ability. Sixteen 18-month-old Yili horses participated in this study. Their heart size and function were measured using ultrasound, while their blood was analyzed to detect changes in fat-related substances. The results showed that training led to noticeable improvements in heart structure. Several types of fats in the blood also changed, mainly those linked to energy production, fat metabolism, and cell health. These findings offer valuable guidance for designing better training programs and ensuring the heart health of sport horses, benefiting both the equestrian industry and animal welfare.

Abstract: The purpose of this study was to explore the relationship between blood lipid levels and the differences in cardiac structure and function of trained and untrained Yili horses as related to exercise performance. We utilized quantitative lipidomics technology to elucidate how the differences in lipid compounds in the blood influenced performance outcomes. Sixteen 18-month-old Yili horses were selected, ten of which received a 15-week training regimen, and six were kept as untrained controls. Cardiac structure and function were assessed by echocardiography, while plasma lipid metabolites were detected and identified by liquid chromatography–mass spectrometry. The results showed that key cardiac structural indices, such as left ventricular end-diastolic diameter, left ventricular end-systolic diameter, and left ventricular posterior wall thickness, were significantly greater in the trained group compared with the untrained group, indicating that exercise training promotes adaptive cardiac remodeling. Regarding lipid metabolites, significant differences were observed between the trained and untrained groups, with a total of 281 lipids identified—212 upregulated and 69 downregulated. These differentially expressed lipids were primarily enriched in pathways such as necroptosis, ether lipid metabolism, and sphingolipid signaling, which are associated with cell migration, survival, proliferation, and regulation of lipid metabolism. Further correlation analysis revealed that differences in certain lipids, such as PE (20:4_18:0), PC (17:0_18:1), and LPC subclasses, were significantly correlated with exercise-mediated cardiac structural and functional changes and exercise performance enhancement. These findings provide novel molecular insights into the effects of exercise training on cardiac structure and lipid metabolism in horses and can serve as a reference for training strategies and preserving cardiac health in performance horses.

Keywords: performance horse training; lipidomics; echocardiography; remodeling; endurance; exercise regimens

1. Introduction

The structure and functionality of a horse's heart can provide it with significant advantages in exercise capacity. The mass of a horse heart is proportionally much greater than that of other species, with an adult horse's heart weighing on average 3.9 kg, while the heart of a racehorse can reach 6.3 kg, and that of a top-performing racehorse can weigh nearly 10 kg [1]. The heart rate of horses can increase from 35 beats per minute to 240 beats per minute, surpassing the capability of most species, including humans, and allowing them to significantly enhance cardiac output in a short time to meet the demands of athletic performance [2]. Researchers have measured heart size using echocardiography and found a significant positive correlation between left ventricular mass and maximal oxygen uptake [3].

Training is a key factor influencing heart size in horses. Studies have shown that high-intensity training leads to cardiac hypertrophy, thereby improving cardiac function. As training intensity increased, the heart mass of mature racehorses, particularly the left ventricle, increased significantly [4]. Compared to horses given lower intensity exercise, those receiving high-intensity training exhibited a larger LVID size [5]. However, excessive cardiac remodeling can also cause myocardial fibrosis, increasing the risk of arrhythmias [6].

The heart consumes a large amount of ATP daily to maintain basal metabolism and normal contraction, which is crucial for sustaining systemic blood pressure. The heart primarily relies on fatty acids as an energy source, with fatty acid oxidation providing most of the ATP required for cardiac contraction [7]. Under varying physiological conditions, the heart can switch metabolic substrates between fatty acids and glucose according to demand, maintaining a delicate balance in cardiac homeostasis through the uptake, storage, and utilization of high-energy fuels. Studies have shown that in patients suffering from heart failure and cardiomyopathy, cardiac energy substrates shift toward glucose utilization [8]. Lipid metabolism and the dynamic alteration of membrane lipid composition are critical for cardiac function. During heart development, an increase in the unsaturation levels of cardiolipin and phospholipids enhances mitochondrial function and energy metabolism [9].

Lipids are essential as both energy sources and basic structural components of the body. In horses, they are influenced by exercise type and intensity and are also vital substrates for skeletal muscle metabolism [10]. In Thoroughbred horses, changes in plasma concentrations of non-esterified fatty acids (NEFAs) and triglycerides before and after intense exercise indicate that lipid metabolism plays a crucial role in skeletal muscle energy production [11]. Lipidomics analysis using liquid chromatography–mass spectrometry (LC/MS) in Thoroughbred horses before and after training identified 933 differential lipids, with the signal intensity of 13 lipids significantly altered due to exercise [12]. A study of Arabian and part-Arabian endurance horses found that lipid biomarkers alone could not accurately determine where a horse would finish in a race, but highly correlated lipid metabolites were identified in the blood of the highest performing horses [13]. A follow-up experiment found that after finishing a race, horses exhibited increased glucose homeostasis, lipid metabolism, ketogenesis, ATP synthesis, and acetate production [14].

There have been numerous studies on exercise metabolomics of performance horses, consistently demonstrating significant differences in blood metabolites before and after exercise [15]. Significant correlations have also been observed between cardiac parameters and exercise performance. However, lipidomics research comparing the types and abundance of plasma lipids in trained vs. untrained horses is limited. This study conducted a comprehensive quantitative lipidomics analysis to examine changes in plasma lipid metabolites in Yili horses after three months of training. It explored the relationship between plasma lipid metabolites and cardiac structure and function and identified key lipid biomarkers distinguishing trained and untrained horses.

2. Materials and Methods

2.1. Ethical Statement

This study was reviewed and approved by the Animal Policy and Welfare Committee of Xinjiang Agricultural University (Ethical Approval No.: 2023037). All experimental procedures strictly followed animal welfare and experimental ethics regulations. Informed consent was obtained from the horse owners. Before the experiment began, all horses underwent veterinary examinations and were confirmed to be in good health with no detected cardiac disease.

2.2. Experimental Design and Horse Grouping

Sixteen 18-month-old Yili horses (each group consisted of an equal number of male and female horses, from a state-owned stud farm in the Yili Kazakh Autonomous Prefecture, Urumqi, China) were selected for this study. The horses had similar birth dates with the same feeding and management conditions (For details, please refer to Supplementary Materials Table S1), minimal body size differences (the body measurement data are similar across groups. For details, please refer to Supplementary Materials Table S2), and were maintained under uniform feeding and management conditions. Ten horses underwent a three-month training program (trained group), while the remaining six were allowed free movement and were not exercised (untrained group). The jockey team at this facility is highly experienced, maintains a similar body weight, and demonstrates excellent professional competence.

2.3. Training Program

The trained group received the following training:

Week 1: desensitization training (both groups participated initially; however, only the training group continued after this week)

Desensitization phase:

1. Teach horse to listen with lead exercises;
2. Practice exercises to make horse comfortable with touch;
3. Train your horse to follow directions through circle work;
4. Introducing your horse to a saddle;
5. Place the saddle on your horse's back;
6. Tighten the girth in intervals;
7. Use a mounting block to climb in the saddle;
8. Mount and dismount from the saddle in 10-min intervals.

Training period, starting from Week 5:

Begin walking your horse while in the saddle.

Week 2: thirty minutes in the lunge ring (walk—5 min, trot—20 min, and walk—5 min);

Week 3: forty minutes in the lunge ring (walk—5 min, trot—30 min, and walk—5 min);

Week 4: one hour on the horse walker (2 m/s—20 min, 3 m/s—30 min, and 4 m/s—10 min);

Week 5: thirty-five minutes of warm-up on the horse walker (2 m/s—5 min, 3 m/s—15 min, and 4 m/s—10 min), fifteen minutes of riding in the training arena (trot—5 minand canter—10 min);

Subsequent three weeks: an additional five minutes of riding (trot—5 min) each week;

Week 9: thirty-five minutes of warm-up on the horse walker (2 m/s—5 min, 3 m/s—15 min, and 4 m/s—15 min), fifteen minutes of track training (trot—5 minand canter—10 min, gallop);

Subsequent three weeks: An additional five minutes of track training per week until the 15th week, marking the end of the training program. (Weeks 10–12: based on Week 9, trot duration increased by 5 min per session. Weeks 13–14: canter duration increased by 5 min per session. Week 15: gallop duration increased by 5 min);

Trot at 50–60% HRmax, canter at 60–70% HRmax, and gallop at 70–80% HRmax.

2.4. Echocardiographic Evaluation

One day after the training ended, echocardiography was performed on the horses in a resting state using a Mindray M6 portable veterinary color Doppler ultrasound system. A 2.5 MHz probe was used for two-dimensional (2D) and M-mode imaging between the third and fourth or fourth and fifth ribs on the right thorax. The maximum imaging depth was set to 30 cm, with the transducer's focal point fixed at 5 cm and a maximum sector angle of 110°. All echocardiographic examinations were conducted by the same operator, and images were recorded only when the heart rate was below 40 beats per minute. Three nonconsecutive cardiac cycles were measured, and the average value was taken.

A total of 22 cardiac parameters were measured and calculated from the right parasternal long-axis, right parasternal short-axis, and M-mode right parasternal short-axis images.

2.5. Quantitative Plasma Lipidomics Analysis by LC/MS-MS

Blood samples (resting state; 10 mL) were collected from the jugular vein of each horse into EDTA anticoagulation tubes. The samples were immediately centrifuged at 15,000× *g* for 15 min, and 1.5 mL aliquots of plasma were transferred into cryotubes, flash-frozen in liquid nitrogen, and stored at −80 °C for later lipidomics analysis.

1. Liquid chromatography column: Thermo Accucore™ C30 column (2.6 μm; 100 mm × 2.1 mm i.d.);
2. Mobile phases:

   Phase A: acetonitrile/water (60/40, *v*/*v*) containing 0.1% formic acid and 10 mmol/L ammonium formate;

   Phase B: acetonitrile/isopropanol (10/90, *v*/*v*) containing 0.1% formic acid and 10 mmol/L ammonium formate;
3. Gradient elution program:

   0 min: A/B = 80:20 (*v*/*v*);
   2 min: A/B = 70:30 (*v*/*v*);
   4 min: A/B = 40:60 (*v*/*v*);
   9 min: A/B = 15:85 (*v*/*v*);
   14 min: A/B = 10:90 (*v*/*v*);
   15.5 min: A/B = 5:95 (*v*/*v*);
   17.3 min: A/B = 5:95 (*v*/*v*);
   17.5 min: A/B = 80:20 (*v*/*v*);
   20 min: A/B = 80:20 (*v*/*v*);
4. Flow rate: 0.35 mL/min;
5. Column temperature: 45 °C;
6. Injection volume: 2 μL.

Lipid identification and quantification were based on the Metware self-built database, MWDB (https://www.metware.cn/, accessed on 29 January 2025). Lipids were identified by measuring their retention time and collecting parent–daughter ion pair data. Pearson correlation analysis was used to determine the relationship between differential metabolites and cardiac structure and function parameters, with the level for significance set at $p < 0.05$.

2.5.1. Data Processing

Mass spectrometry data were processed using Analyst 1.6.3 software, and metabolites were identified and quantified using a local metabolic database. Detectable ions were screened using a triple quadrupole mass spectrometer, and their signal intensity (CPS) was recorded. Chromatographic peak integration and correction were performed using MultiQuant software 3.0.3. To ensure data reproducibility, a quality control (QC) sample was inserted after every ten samples. The final dataset was simplified and reduced through multivariate statistical analysis.

Principal component analysis (PCA) was conducted using the prcomp function in R software (version 3.5.1) with scale = true for unit variance scaling to identify metabolic differences between samples. Hierarchical clustering analysis (HCA) of metabolite data processed with unit variance scaling was performed using the ComplexHeatmap package in R, generating heatmaps to visualize the metabolic accumulation patterns across different samples.

Differential metabolites were screened using both univariate and multivariate analysis methods, including hypothesis testing, fold change (FC) analysis, and orthogonal partial least squares discriminant analysis (OPLS-DA). The selection criteria were VIP > 1 and $p < 0.05$, identifying metabolites significantly associated with training status.

2.5.2. Screening of Differential Metabolites and KEGG Functional Analysis

The identified differential metabolites were functionally annotated using the KEGG database to identify associated biological pathways. Volcano plots were generated using the ggplot2 package in R to visualize significantly different metabolites between groups, while the pheatmap package was used to generate heatmaps showing metabolite clustering across samples. KEGG enrichment analysis was conducted to assess pathway significance, with a screening threshold of $p < 0.05$, identifying significantly enriched metabolic pathways.

2.5.3. Correlation Analysis Between Cardiac Structure, Function, and Plasma Lipid Composition

Pearson correlation analysis in SPSS 26.0 was used to assess the relationships between cardiac structure and function indices and 153 differential lipid metabolites with VIP > 1.5. The significance thresholds were set at $p < 0.05$ (*) for significant correlations, $p < 0.01$ (**) for highly significant correlations, and $p < 0.001$ (***) for very highly significant correlations.

2.6. Statistical Analysis

All figures were generated using GraphPad Prism 8.0 (GraphPad Software Inc., San Diego, CA, USA). Statistical analyses included analysis of variance (ANOVA), conducted using SPSS 26.0 (IBM, Armonk, NY, USA). Data are expressed as mean ± standard deviation, and group differences were assessed using one-way ANOVA. Homogeneity of variance within groups was tested, with $p > 0.05$ indicating no significant variance difference.

3. Results

3.1. Analysis of Differences in Cardiac Structure Between Trained and Untrained Yili Horses

This study included 16 Yili horses, with 10 horses undergoing three months of training compared with 6 untrained horses moving freely in the activity field under the same feeding and management conditions. Table 1 summarizes the cardiac structure and functional characteristics of all horses. Figure 1a shows there are no significant differences in ejection fraction (EF%) and fractional shortening (FS%) between the trained and untrained groups. However, significant differences were observed in other cardiac structural and functional parameters (Figure 1a,b).

Table 1. Results of echocardiography of the two groups of Yili horses under resting conditions showing cardiac structural and functional measurements.

TERM	Untrained Group (UG)	Trained Group (TG)
RVDd (cm)	2.17 ± 0.09	2.37 ± 0.13
IVSd (cm)	1.97 ± 0.19	2.05 ± 0.18
LVIDd (cm)	8.61 ± 0.21	9.09 ± 0.26
LVFWd (cm)	1.48 ± 0.07	1.59 ± 0.13
RVDs (cm)	1.44 ± 0.16	1.47 ± 0.1
IVSs (cm)	3.21 ± 0.35	3.58 ± 0.21
LVIDs (cm)	5.14 ± 0.07	5.39 ± 0.1
LVFWs (cm)	2.43 ± 0.22	2.71 ± 0.31
LV minor (cm)	13.48 ± 0.7	14.22 ± 0.43
LADd (cm)	7.78 ± 0.55	8.23 ± 0.29
LADs (cm)	8.88 ± 0.3	9.77 ± 0.51
AODd (cm)	4.39 ± 0.37	4.53 ± 0.22
PAd (cm)	3.56 ± 0.16	3.67 ± 0.17
AODs (cm)	4.61 ± 0.23	5.11 ± 0.18
PAs (cm)	4.11 ± 0.31	4.27 ± 0.17
EF (%)	0.69 ± 0.01	0.69 ± 0.01
FS%	0.4 ± 0.01	0.41 ± 0.01
LVM (kg)	1.14 ± 0.09	1.36 ± 0.19
EDV (mL)	405.77 ± 21.99	457.75 ± 29.38
ESV (mL)	125.85 ± 3.96	140.92 ± 6.41
SV (mL)	279.92 ± 18.39	316.83 ± 23.98
HR	45.03 ± 2.24	43.93 ± 3.96
CO (L/min)	15.42 + 1.41	13.89 + 1.27

Note: UG, untrained control group, $n = 6$; TG, trained group, $n = 10$; RVDd: end-diastolic right ventricular diameter; IVSd: end-diastolic interventricular septal thickness; LVIDd: end-diastolic left ventricular diameter; LVFWd: end-diastolic left ventricular free wall thickness; LADd: end-diastolic left atrial diameter; LADs: end-systolic left atrial diameter; AODd: end-diastolic aortic root diameter; PADd: end-diastolic pulmonary artery diameter; PADs: end-systolic pulmonary artery diameter; EF: ejection fraction; FS: fractional shortening; LVminor: left ventricular minor; SV: stroke volume; EDV: end-diastolic left ventricular volume; ESV: end-systolic left ventricular volume; CO: cardiac output; LVM: left ventricular myocardial mass; HR: heart rate. Data are reported as mean + SD for all variables.

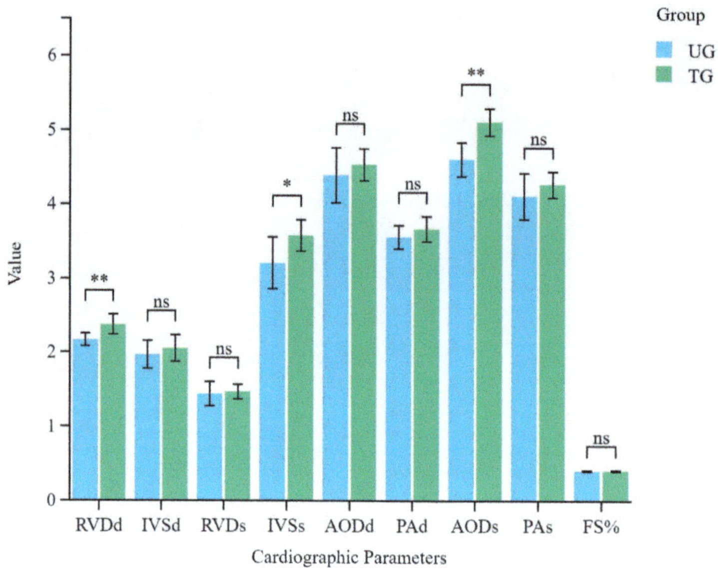

(a)

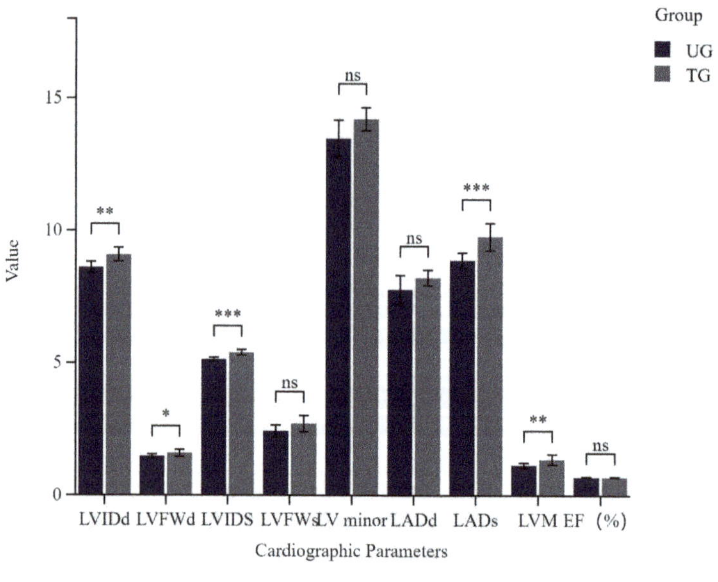

(b)

Figure 1. Statistical comparisons of echocardiography results from the two groups(UG and TG) of Yili horses. (**a**) Differences of RVDd, IVSd, RVDs, IVSs, AODd, PAd, AODs, PAs, FS% between UG group and TG group. (**b**) Differences of LVIDd, LVFWd, LVIDs, LVFWs, LV minor, LADd, LADs, LVM, EF% between UG group and TG group. ns: not significant, * $p < 0.05$, ** $p < 0.01$, and *** $p < 0.001$.

3.2. Changes in Quantitative Lipidomics Between Trained and Untrained Yili Horses

3.2.1. Identification of Differentially Expressed Lipid Metabolites in Plasma

Using lipidomics technology and UPLC-MS/MS quantitative data, 347 metabolites were identified. Principal component analysis (PCA) was performed on all experimental samples and quality control (QC) samples to preliminarily assess the degree of variation between samples (Figure 2a). The analysis revealed differences among samples, with distinct separation observed between the trained group (TG) and the untrained group (UG). All samples fell within the elliptical confidence interval, indicating clear differentiation between the two groups.

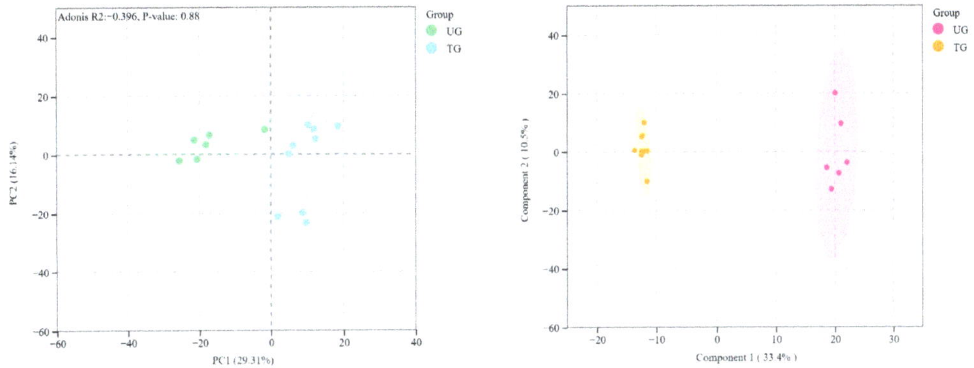

(**a**) (**b**)

Figure 2. Quality control analysis of equine plasma in UG versus TG. (**a**) Plot of principal component analysis. (**b**) Map of OPLS-DA scores.

Orthogonal partial least squares discriminant analysis (OPLS-DA) was used to analyze the lipid metabolites in the plasma of Yili horses before and after training. The score plot showed a clear separation between the trained and untrained Yili horses after three months of training, demonstrating significant changes in plasma lipid metabolites (Figure 2b).

3.2.2. Analysis of Differentially Expressed Lipid Metabolites in Plasma

To gain a clearer understanding of the lipid metabolic changes between trained and untrained Yili horses, we used an absolute quantitative lipidomics approach to identify lipid subclasses and lipid molecules in the samples. A total of 34 lipid subclasses and 281 lipid molecules were detected in horse plasma (Figure 3a). Compared with untrained Yili horses, 212 lipids were upregulated, and 69 lipids were downregulated in the trained group. This indicates that exercise training plays a crucial role in the regulation of lipid metabolism, suggesting that it could be used for improving lipid metabolic health. Interestingly, although a larger number of lipid metabolites were classified under the glycerophospholipid (GP) category, the triacylglycerol (TG) subclass within the glycerolipid (GL) category showed a numerical advantage (Figure 3b).

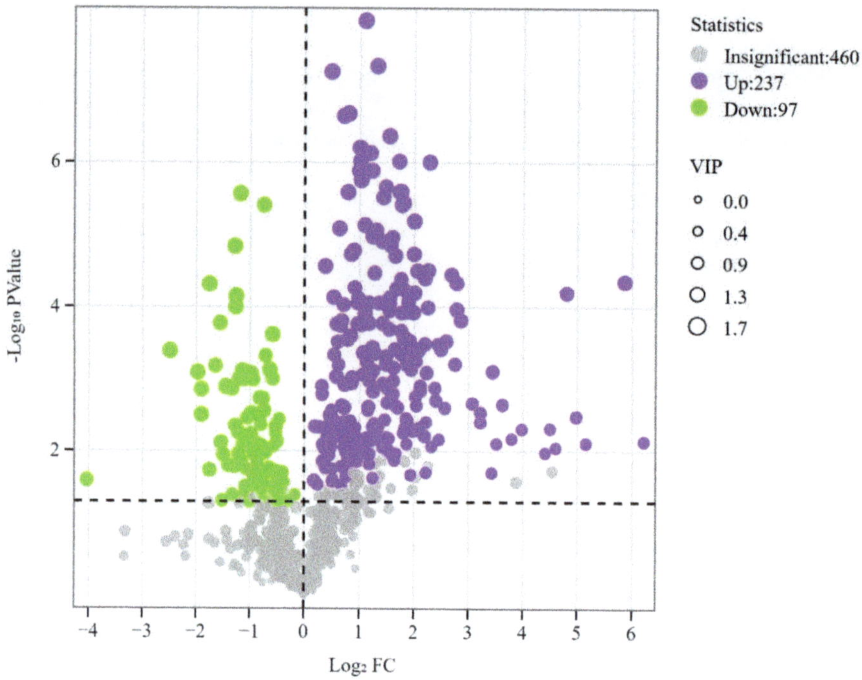

(a)

(b)

Figure 3. Metabolome analysis of equine plasma in UG and TG. (**a**) Volcano plot of differential lipids between trained and untrained Yili horses. (**b**) Bar chart of the number of differential lipid subclasses between trained and untrained Yili horses.

3.2.3. Pathway Enrichment Analysis of Differential Lipid Metabolites

All differential lipid metabolites between trained and untrained Yili horses were mapped to the KEGG database to obtain information on the metabolic pathways in which these metabolites participate. An enrichment analysis was performed on the annotated results to identify pathways with the highest enrichment of differential metabolites, and a KEGG enrichment bubble plot was generated (Figure 4). The differential lipids between trained and untrained Yili horses were primarily annotated and enriched in pathways such as necroptosis, ether lipid metabolism, and the sphingolipid signaling pathway, which are involved in cell migration, survival, and proliferation. Additionally, significant enrichment was observed in pathways related to linoleic acid metabolism, arachidonic acid metabolism, and alpha-linolenic acid metabolism.

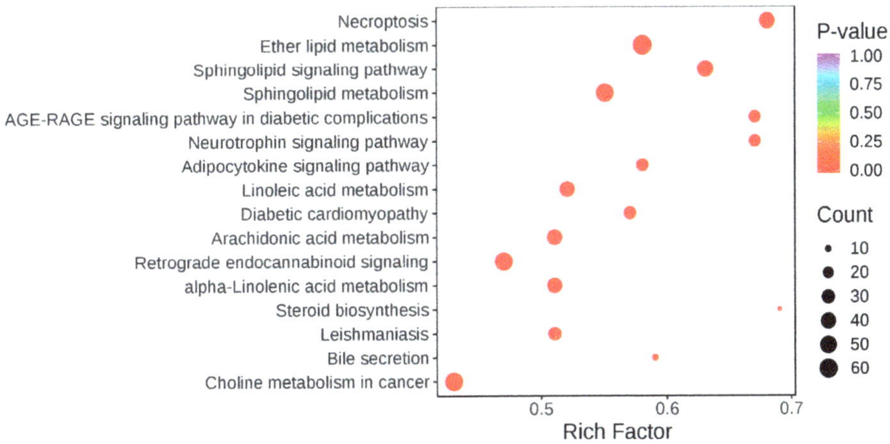

Figure 4. KEGG enrichment bubble plot of differential lipids between the trained and untrained groups. p-value: significance test p-value; count: number of lipids enriched in the pathway; rich factor: enrichment factor, representing the ratio of the number of differential lipids annotated to a given pathway to the total number of lipids annotated to that pathway. A higher rich factor indicates a greater degree of enrichment.

3.2.4. Correlation Analysis Between Differential Metabolites and Cardiac Structure

As shown in Figure 5, the variable importance in projection (VIP) value represents the influence strength of each lipid's intergroup differences in classifying samples within the model. A higher VIP value indicates that the corresponding lipid contributes more significantly to the differentiation between the two groups. By performing Pearson correlation analysis between key metabolites with high VIP values and cardiac structure and function, we identified significant correlations between them.

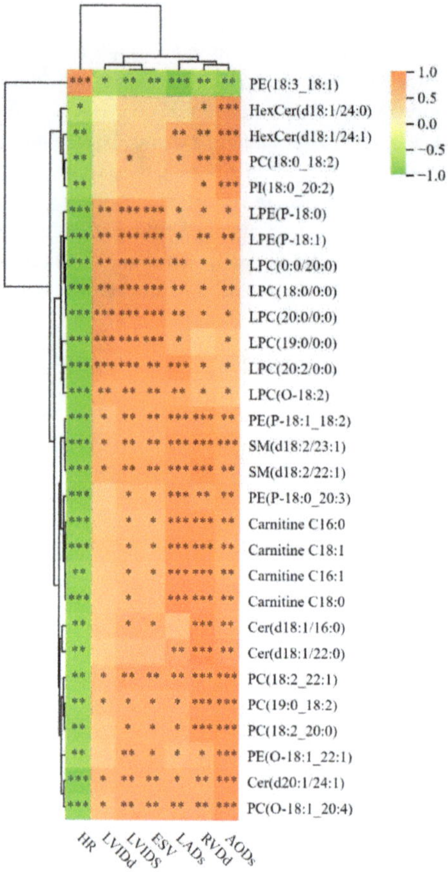

Figure 5. Correlation between lipid metabolites and cardiac structure and function. * $p < 0.05$, ** $p < 0.01$, and *** $p < 0.001$. Orange indicates a positive correlation, while green indicates a negative correlation.

4. Discussion

Cardiac remodeling refers to compensatory or decompensatory changes in cardiac genes, proteins, cells, and the intercellular matrix, which clinically manifest as structural and functional alterations of the heart [16]. Lipid metabolism is an indispensable component of cardiac remodeling, as lipids serve as the major energy source for the heart and are influenced by multiple factors [17]. This study compared the differences in cardiac structure, function, and absolute quantitative lipids in the plasma between trained and untrained Yili horses, providing new insights into the molecular mechanisms of physiological cardiac remodeling induced by long-term training.

The Yili horse is a breed native to the Xinjiang Uygur Autonomous Region of China. They combine the advantages of their Kazakh mare lineage, such as the ability to digest rough forage and strong disease resistance, with the speed and endurance of Thoroughbred sires. Although the statistics are incomplete, more than 16,000 Yili horses with athletic potential have been registered with the Xinjiang Horse Industry Association (https://horse.xjau.edu.cn/, accessed on 29 January 2025).

The heart of a horse accounts for approximately 0.9% to 1% of its total body mass, and in systematically trained horses, heart mass can reach 1.1% of total body weight [18,19].

The left ventricle (LV) is the primary pumping chamber, increasing myocardial contractility and cardiac output during exercise to meet the body's oxygen and nutrient demands. After prolonged racing and systematic training, myocardial cells enlarge, leading to adaptive left ventricular hypertrophy, which significantly increases total heart mass [4,5,20]. Our echocardiographic measurements revealed that LVIDd, LVIDs, LVFWd, and IVSs were significantly greater in the trained group compared to the untrained group, while LVFWs, IVSd, and LV minor showed no statistical difference but were still larger in the trained group. A review of exercise-induced cardiac remodeling concluded that a wealth of literature supported the notion that cardiac remodeling was a widespread phenomenon, with the distinctive characteristics of an "athlete's heart" observed in humans, horses, and dogs [1]. At rest, trained horses had significantly lower heart rates than untrained horses, while EDV, ESV, SV, CO, and LVM were significantly higher, indicating that three months of incremental training contributes to adaptive cardiac remodeling.

Most studies on equine cardiac remodeling focused on the left ventricle due to its crucial role in maintaining cardiac output during exercise. However, recent research has revealed a dynamic interdependence between the right and left ventricles. The right ventricle supports left ventricular filling by increasing end-diastolic volume, thereby maintaining overall cardiac pump function [21]. Another study found that different athletic disciplines only influenced right ventricular structure in 35% of athletes, with changes being non-significant [22]. In this study, RVDd was significantly increased in the trained group compared to the untrained group, while RVDs showed no significant changes, suggesting that longer training durations may be required for substantial right ventricular adaptation. The aorta plays a crucial role in buffering pulsatile cardiac output to reduce the risk of vascular damage due to blood pressure fluctuations [23]. Long-term training induces adaptive changes in the aorta, which contribute to cardiovascular health [24]. In this study, AODs was significantly larger in the trained group, whereas AODd did not follow the same trend, indicating that training positively stimulated aortic structural adaptation in Yili horses.

The relationship between lipids and physiological cardiac remodeling is complex and multifaceted. Lipids play a key role in cardiac metabolism and energy production, and their dysregulation can significantly alter cardiac function and structure. In exercise-induced cardiac remodeling, lipids contribute to metabolic adaptations that support increased cardiac workload capability and efficiency [25].

Phosphatidylcholine (PC) and phosphatidylethanolamine (PE) are the most abundant phospholipids in cell membranes. PC not only maintains mitochondrial membrane integrity but also regulates mitochondrial bioenergetics, supporting efficient electron transport chain function [26]. Another study found that the peak values of PC (20:4/18:0) and PE (20:4/18:0) were negatively correlated with the heart weight/body weight ratio, suggesting that these lipids may have protective roles in cardiac remodeling [27]. Lysophosphatidylcholine (LPC), a hydrolysis product of PC, has important biological functions. LPC is involved in the pathogenesis of cardiovascular diseases, including atherosclerosis and inflammatory responses [28]. In this study, PC (17:0_18:1) (VIP = 1.98) and PE (20:4_18:0) (VIP = 2.01) showed significant differences between the two groups and strong positive correlations with RVDd and AODs. Additionally, LPC subclasses (LPC (18:0/0:0), LPC (0:0/15:0), LPC (15:0/0:0), and LPC (20:0/0:0)) were significantly correlated with LVID, further emphasizing the role of lipid metabolism in cardiac structural adaptations.

Exercise can regulate lipid metabolism by influencing intracellular ceramide (Cer) levels, which are essential for energy balance and metabolic homeostasis. [29]. Ceramide plays a key role in muscle energy metabolism, and its level is correlated with metabolic state and energy demand. Elevated ceramide levels can increase muscle cell apoptosis, affecting muscle mass and function [30]. This study found that Cer (d20:1/24:1) was more abundant in the trained group and significantly positively correlated with AODs.

The differential lipids between trained and untrained horses were enriched in the sphingolipid signaling pathway. Sphingomyelin (SM) was not only widely present in neuronal cell membranes but also regulates signal transduction and cellular metabolism. Studies have shown that sphingomyelin influenced membrane protein function and interactions, playing a key role in neuronal signaling and intercellular communication [31,32]. SM (d40:4) was primarily enriched in the sphingolipid signaling pathway, which has been shown to be essential for myocardial proliferation and cardiac regeneration [33]. Linoleic acid metabolism is vital for cardiovascular health, as linoleic acid derivatives regulate inflammation and vascular function, while arachidonic acid metabolism influences inflammatory responses, vascular tone, and platelet aggregation [34,35]. These enrichments of metabolic and signaling pathways demonstrate that intensive exercise training can significantly enhance cardiac energy metabolism, structural adaptation, and functional optimization. This study provides new theoretical support for understanding the adaptive mechanisms of exercise training on cardiac function and for designing more effective exercise regimens.

5. Conclusions

In summary, the differential plasma lipids PE (20:4_18:0), PC (17:0_18:1), and LPC subclasses in trained and untrained Yili horses were significantly correlated with cardiac structure and function. The trained group exhibited more pronounced cardiac remodeling and metabolic regulation advantages, including enhanced metabolic adaptation and structure–function optimization through pathways such as the sphingolipid signaling pathway, linoleic acid metabolism, and arachidonic acid metabolism. This study integrated cardiac structure, function, and lipid metabolism to evaluate the differences between trained and untrained Yili horses, potentially providing insights into more effective training strategies and cardiac regulation mechanisms for sport horses.

Supplementary Materials: The following supporting information can be downloaded at https://www.mdpi.com/article/10.3390/vetsci12030255/s1: Table S1: Composition and nutritional levels of the basal diet (on dry matter basis); Table S2: Body measurements of horses before training.

Author Contributions: T.W.: Investigation, Software, Project administration, Writing—Original Draft, Writing—Review and Editing; J.M.: Methodology; J.W.: Methodology, Writing—Review and Editing; W.R.: Methodology, Writing—Review and Editing; W.A.: Validation; Formal analysis; Y.B.: Resources; X.Y. (Xixi Yang): Data Curation; Y.Z.: Conceptualization, Supervision, Funding acquisition; X.Y. (Xinkui Yao): Conceptualization, Supervision, Funding acquisition. All authors have read and agreed to the published version of the manuscript.

Funding: This study was supported by the National Natural Science Foundation of China Youth Program (grant No. 32202667) and the Major Science and Technology Project of Xinjiang Uygur Autonomous Region (grant No. 2022A02013-1); Key Laboratory Project of Xinjiang Horse Breeding and Exercise Physiology (grant No. 2521KJTXM); Xinjiang Uygur Autonomous Region Natural Science Foundation Youth Science Fund (grant No. 2024D01B40); and Central Guidance for Local Science and Technology Development Fund (grant No. ZYYD2025JD02).

Institutional Review Board Statement: The experimental procedures adhered to the ARRIVE guidelines and were approved by the Animal Welfare Committee of Xinjiang Agricultural University (Approval No. 2023037). The owners of all horses provided informed consent for the experiments.

Informed Consent Statement: Informed consent was obtained from all subjects involved in the study.

Data Availability Statement: The data supporting this study's findings are available from the corresponding author upon reasonable request.

Conflicts of Interest: The authors declare that they have no competing interests.

References

1. Shave, R.; Howatson, G.; Dickson, D.; Young, L. Exercise-induced cardiac remodeling: Lessons from humans, horses, and dogs. *Vet. Sci.* **2017**, *4*, 9. [CrossRef] [PubMed]
2. Higgins, G. *Horse Anatomy for Performance: A Practical Guide to Training, Riding and Horse Care*; David & Charles: Exeter, UK, 2012.
3. Vernemmen, I.; Vera, L.; Van Steenkiste, G.; van Loon, G.; Decloedt, A. Reference values for 2-dimensional and M-mode echocardiography in Friesian and Warmblood horses. *J. Vet. Intern. Med.* **2020**, *34*, 2701–2709. [CrossRef]
4. Pinar, O.; Sancak, A. Effects of different heart dimensions on race performance in Thorougbred race horses. *Acta Sci. Vet.* **2018**, 46. [CrossRef]
5. Buhl, R.; Ersbøll, A.K.; Eriksen, L.; Koch, J. Changes over time in echocardiographic measurements in young Standardbred racehorses undergoing training and racing and association with racing performance. *J. Am. Vet. Med. Assoc.* **2005**, *226*, 1881–1887. [CrossRef] [PubMed]
6. Nath, L.; Saljic, A.; Buhl, R.; Elliott, A.; La Gerche, A.; Ye, C.; Schmidt Royal, H.; Lundgren Virklund, K.; Agbaedeng, T.; Stent, A. Histological evaluation of cardiac remodelling in equine athletes. *Sci. Rep.* **2024**, *14*, 16709. [CrossRef]
7. Kadkhodayan, A.; Coggan, A.R.; Peterson, L.R. A "PET" area of interest: Myocardial metabolism in human systolic heart failure. *Heart Fail. Rev.* **2013**, *18*, 567–574. [CrossRef]
8. Actis Dato, V.; Lange, S.; Cho, Y. Metabolic Flexibility of the Heart: The Role of Fatty Acid Metabolism in Health, Heart Failure, and Cardiometabolic Diseases. *Int. J. Mol. Sci.* **2024**, *25*, 1211. [CrossRef]
9. Miao, H.; Li, B.; Wang, Z.; Mu, J.; Tian, Y.; Jiang, B.; Zhang, S.; Gong, X.; Shui, G.; Lam, S.M. Lipidome atlas of the developing heart uncovers dynamic membrane lipid attributes underlying cardiac structural and metabolic maturation. *Research* **2022**, *2022*, 0006. [CrossRef]
10. Jang, J.H.; Kim, K.D.; Kim, K.K.; Park, P.J.; Choi, C.J.; Oh, O.J.; Song, S.K.; Kim, K.S.; Cho, C.B. Analysis of metabolomic patterns in thoroughbreds before and after exercise. *Asian-Australas. J. Anim. Sci.* **2017**, *30*, 1633–1642. [CrossRef]
11. Piccione, G.; Arfuso, F.; Fazio, F.; Bazzano, M.; Giannetto, C. Serum lipid modification related to exercise and polyunsaturated fatty acid supplementation in jumpers and thoroughbred horses. *J. Equine Vet. Sci.* **2014**, *34*, 1181–1187. [CrossRef]
12. Nolazco Sassot, L.; Villarino, N.; Dasgupta, N.; Morrison, J.; Bayly, W.; Gang, D.; Sanz, M. The lipidome of Thoroughbred racehorses before and after supramaximal exercise. *Equine Vet. J.* **2019**, *51*, 696–700. [CrossRef] [PubMed]
13. Le Moyec, L.; Robert, C.; Triba, M.N.; Billat, V.L.; Mata, X.; Schibler, L.; Barrey, E. Protein catabolism and high lipid metabolism associated with long-distance exercise are revealed by plasma NMR metabolomics in endurance horses. *PLoS ONE* **2014**, *9*, e90730. [CrossRef] [PubMed]
14. Le Moyec, L.; Robert, C.; Triba, M.N.; Bouchemal, N.; Mach, N.; Riviere, J.; Zalachas-Rebours, E.; Barrey, E. A first step toward unraveling the energy metabolism in endurance horses: Comparison of plasma nuclear magnetic resonance metabolomic profiles before and after different endurance race distances. *Front. Mol. Biosci.* **2019**, *6*, 45. [CrossRef] [PubMed]
15. Wang, T.; Zeng, Y.; Ma, C.; Meng, J.; Wang, J.; Ren, W.; Wang, C.; Yuan, X.; Yang, X.; Yao, X. Plasma Non-targeted Metabolomics Analysis of Yili Horses Raced on Tracks with Different Surface Hardness. *J. Equine Vet. Sci.* **2023**, *121*, 104197. [CrossRef]
16. Schirone, L.; Forte, M.; Palmerio, S.; Yee, D.; Nocella, C.; Angelini, F.; Pagano, F.; Schiavon, S.; Bordin, A.; Carrizzo, A. A review of the molecular mechanisms underlying the development and progression of cardiac remodeling. *Oxidative Med. Cell. Longev.* **2017**, *2017*, 3920195. [CrossRef]
17. Lopaschuk, G.D.; Ussher, J.R.; Folmes, C.D.; Jaswal, J.S.; Stanley, W.C. Myocardial fatty acid metabolism in health and disease. *Physiol. Rev.* **2010**, *90*, 207–258. [CrossRef]
18. Hanson, C.; Kline, K.; Foreman, J. Measurements of heart scores and heart weights in horses of two different morphic body types. *Comp. Biochem. Physiol. Part A Physiol.* **1994**, *108*, 175–178. [CrossRef]
19. José Lopes, V.; Barbosa da Costa, G.; Garcia Pereira, A.; de Freitas Pereira, L.; Fernandes Magalhães, L.; dos Santos, D.; Paulino Junior, D. Morphoquantitative evaluation of the heart of equine athletes. *Pubvet* **2021**, *15*, a826.
20. Young, L.E. Cardiac responses to training in 2-year-old thoroughbreds: An echocardiographic study. *Equine Vet. J.* **1999**, *31*, 195–198. [CrossRef]
21. Naeije, R.; Badagliacca, R. The overloaded right heart and ventricular interdependence. *Cardiovasc. Res.* **2017**, *113*, 1474–1485. [CrossRef]
22. Maskhuliva, L.; Akhalkatsi, V.; Chelidze, K.; Kakhabrishvili, Z.; Matiashvili, M.; Chabashvili, N.; Chutkerashvili, T. Echocardiographic Study of Right Ventricular Remodeling in Top-Level Georgian Athletes. *Br. J. Sports Med.* **2014**, *48*, 634. [CrossRef]

23. Jacob, M.; Chappell, D.; Becker, B.F. Regulation of blood flow and volume exchange across the microcirculation. *Crit. Care* **2016**, *20*, 1–13. [CrossRef]
24. Morra, E.A.; Zaniqueli, D.; Rodrigues, S.L.; El-Aouar, L.M.; Lunz, W.; Mill, J.G.; Carletti, L. Long-term intense resistance training in men is associated with preserved cardiac structure/function, decreased aortic stiffness, and lower central augmentation pressure. *J. Hypertens.* **2014**, *32*, 286–293. [CrossRef] [PubMed]
25. Gibb, A.A.; Hill, B.G. Metabolic coordination of physiological and pathological cardiac remodeling. *Circ. Res.* **2018**, *123*, 107–128. [CrossRef]
26. Schugar, R.C.; Gliniak, C.M.; Osborn, L.J.; Massey, W.; Sangwan, N.; Horak, A.; Banerjee, R.; Orabi, D.; Helsley, R.N.; Brown, A.L. Gut microbe-targeted choline trimethylamine lyase inhibition improves obesity via rewiring of host circadian rhythms. *eLife* **2022**, *11*, e63998. [CrossRef] [PubMed]
27. Dong, S.; Zhang, R.; Liang, Y.; Shi, J.; Li, J.; Shang, F.; Mao, X.; Sun, J. Changes of myocardial lipidomics profiling in a rat model of diabetic cardiomyopathy using UPLC/Q-TOF/MS analysis. *Diabetol. Metab. Syndr.* **2017**, *9*, 1–9. [CrossRef]
28. Drechsler, R.; Chen, S.-W.; Dancy, B.C.; Mehrabkhani, L.; Olsen, C.P. HPLC-based mass spectrometry characterizes the phospholipid alterations in ether-linked lipid deficiency models following oxidative stress. *PLoS ONE* **2016**, *11*, e0167229. [CrossRef]
29. Tanaka, T.; Nishimura, A.; Nishiyama, K.; Goto, T.; Numaga-Tomita, T.; Nishida, M. Mitochondrial dynamics in exercise physiology. *Pflügers Arch.-Eur. J. Physiol.* **2020**, *472*, 137–153. [CrossRef]
30. Amorese, A.J.; Ryan, A.S. Home-based tele-exercise in musculoskeletal conditions and chronic disease: A literature review. *Front. Rehabil. Sci.* **2022**, *3*, 811465. [CrossRef]
31. Feng, L.T.; Chen, Z.N.; Bian, H. Skeletal muscle: Molecular structure, myogenesis, biological functions, and diseases. *MedComm* **2024**, *5*, e649. [CrossRef]
32. Da, Y.S.; Takagi, H.; Hiroshima, M.; Matsuoka, S.; Ueda, M. Sphingomyelin metabolism underlies Ras excitability for efficient cell migration and chemotaxis. *Cell Struct. Funct.* **2023**, *48*, 145–160.
33. Ji, X.; Chen, Z.; Wang, Q.; Li, B.; Wei, Y.; Li, Y.; Lin, J.; Cheng, W.; Guo, Y.; Wu, S. Sphingolipid metabolism controls mammalian heart regeneration. *Cell Metab.* **2024**, *36*, 839–856.e8. [CrossRef] [PubMed]
34. Thompson, M.; Ulu, A.; Mukherjee, M.; Yuil-Valdes, A.G.; Thoene, M.; Van Ormer, M.; Slotkowski, R.; Mauch, T.; Anderson-Berry, A.; Hanson, C.K. Something Smells Fishy: How Lipid Mediators Impact the Maternal–Fetal Interface and Neonatal Development. *Biomedicines* **2023**, *11*, 171. [CrossRef] [PubMed]
35. Djuricic, I.; Calder, P.C. Beneficial outcomes of omega-6 and omega-3 polyunsaturated fatty acids on human health: An update for 2021. *Nutrients* **2021**, *13*, 2421. [CrossRef]

veterinary sciences

Review

The Genomic Characterization of Equid Alphaherpesviruses: Structure, Function, and Genetic Similarity

Diqiu Liu *,†, Xiaoyang Zhao † and Xiaojun Wang *

State Key Laboratory for Animal Disease Control and Prevention, Harbin Veterinary Research Institute, Chinese Academy of Agricultural Sciences, Harbin 150069, China; zhaoxiaoyang2412@163.com
* Correspondence: liudiqiu@caas.cn (D.L.); wangxiaojun@caas.cn (X.W.)
† These authors contributed equally to this work.

Simple Summary: Genomic structure, genomic function, and genetic similarity are fundamental components of the field of genomics. Equid herpesviruses in subfamily alphaherpesvirinae include equine herpesvirus 1 (EHV-1), EHV-3, EHV-4, EHV-6, EHV-8, and EHV-9, which are recognized as etiological agents responsible for respiratory, urogenital, and neurological disorders in equine species, exhibiting unique and similar characteristics in infection that are influenced by both the identities and differences among their respective genomic homologs. This review systematically examined and synthesized the existing genomic knowledge on αEHVs, focusing on genomic structure, function, and genetic similarity, and conducted pairwise alignments between each homolog. Furthermore, this study identifies essential challenges encountered during the research process and proposes potential solutions. In the current context, future research should prioritize the exploration of unknown genomic functions and novel transcripts within the αEHV genome.

Abstract: Equine herpesvirus 1 (EHV-1), EHV-4, EHV-8, and EHV-9, are classified within the subfamily *Alphaherpesvirinae* and are recognized as causative agents of respiratory, urogenital, and neurological disorders in horses. These viruses, collectively referred to as αEHVs, exhibits both unique and shared characteristics in terms of host interaction, pathogenesis, epidemiology, and immune evasion, which arise from both the identities and discrepancies among respective genomic homologs. The genomic architecture of αEHVs is similar to other members of the same subfamily, such as well-known HSV-1, VZV, and PRV. However, research on the molecular mechanisms underlying αEHV infection and immune response remains significantly less advanced compared to studies on human, porcine, and bovine herpesviruses. This paper systematically describes the genomic structure, function, and genetic similarities of αEHVs and conducts a comparative analysis of selected αEHVs through pairwise sequence alignments of nucleotides and amino acids. This review offers an extensive synthesis of the current understanding related to the study of αEHVs, highlighting the challenges and potential solutions for future research endeavors.

Keywords: equid alphaherpesvirus; genomic structure; function; comparative analysis; future research

1. Introduction

The order *Herpesvirales* consists of families *Orthoherpesviridae*, *Alloherpesviridae*, and *Malacoherpesviridae*, each comprising viruses associated with distinct hosts. Members of *Herpesviridae* infect mammals, birds, or reptiles, while those of *Alloherpesviridae* infect fish or frogs. *Malacoherpesviridae* viruses infect mollusks [1]. *Orthoherpesviridae* has three subfamilies: *Alphaherpesvirinae*, *Betaherpesvirinae*, and *Gammaherpesvirinae* [2]. The five genera of *Alphaherpesvirinae*, including *Iltovirus*, *Mardivirus*, *Scutavirus*, *Simplex virus*, and *Varicellovirus*, comprise all known equid alphaherpesviruses (αEHVs): EHV-1, EHV-3, EHV-4, EHV-6, EHV-8, and EHV-9. EHV-1 shares a close evolutionary relationship and genomic organization with EHV-3 [3]. EHV-3 is the etiological agent responsible for equine coital exanthema, a typically mild genital infection in horses that is transmitted venereally through semen; however, it has been nearly eradicated [4]. EHV-6 is closely related to EHV-3 from the aspect of equine coital exanthema [5]. EHV-6 and EHV-9 are primarily associated with infections in asses and zebras [6]. This review concentrates on the most prevalent and significant αEHVs (EHV-1, EHV-4, and EHV-8) affecting domestic horses and donkeys [6], which are linked to abortion, respiratory diseases, and neurological disorders.

EHV-1 infection can cause upper respiratory disease in horses, presenting symptoms akin to those caused by equine influenza virus [7,8]. In addition to respiratory symptoms, EHV-1 frequently induces neurological syndromes, including ataxia, muscle paresis, paralysis, bladder atony, recumbency, or death. Additionally, EHV-1 has been associated with uterine or placental damage during early gestation and abortion in the late gestation of mares. Horses across all ages occasionally display the classical sings of myeloencephalitis [9]. Similarly, EHV-4 can cause upper respiratory infections in young horses and can also lead to encephalomyelitis with occasional cases of abortion reported [10,11]. Although these two viruses are closely related, they are distinct entities. They were reclassified by the International Committee on Taxonomy of Viruses [1,12]. Infection by EHV-4 typically stays confined to the upper respiratory system, while EHV-1 infection may become systemic following viremia associated with leukocytes [13].

EHV-8, originally designated as asinine herpesvirus 3 (AHV-3) [14], has been implicated in causing febrile rhinitis in donkeys [15–17] and abortion in mares [18]. It is noteworthy that the EHV-8 wh strain was isolated from a horse exhibiting neurological symptoms [19]. Phylogenetically, EHV-8 exhibits a closer relationship to EHV-1 and EHV-9 than to EHV-4 [18–21].

EHV-9, initially termed gazelle herpesvirus-1, represents the most recent addition to the αEHVs group and was first isolated from Thomson's gazelles that succumbed to fulminant encephalitis [22,23]. It is closely related to EHV-1 and EHV-8 and exhibits a broad host range encompassing species such as hamsters, goats, dogs, cats, polar bears, zebras, giraffes, and gazelles, with the exception of those exhibiting lethal encephalitis [24–29]. In equines, however, the clinical presentation is comparatively mild, characterized by moderate encephalitis with reduced neuronal loss, perivascular cuffing, and gliosis [22].

The primary entry point for these viruses is the mucosal surfaces of the respiratory and urogenital tracts, where they initially manifest clinically and often lead to secondary infections. αEHVs are able to establish lifelong latent infections within the lymph nodes, trigeminal ganglia, and leukocytes; however, the exact primary site of latency remains a topic of ongoing debate. Currently, there is no conclusive evidence regarding the incidence and significance of reactivation from latency of EHV-1 and EHV-4. Nonetheless, anecdotal reports suggest that stressful events such as weaning, castration, or transportation may serve as potential causal factors.

This review offers a systematic summary of the current understanding of the genomic structure, function, and genetic similarity of αEHVs and conducts a comparative analysis through pairwise alignment among EHV-1, EHV-4, EHV-8, and EHV-9. These analyses may facilitate future research endeavors aimed at elucidating the molecular mechanisms underlying the distinct and shared characteristics associated with host interaction, pathogenesis, epidemiology, and immune evasion, as influenced by genetic diversity and identity.

2. Genomic Structure

The GenBank accession numbers corresponding to the EHV strains analyzed in this study include EHV-1 strain Ab4 (AY665713), EHV-4 strain NS80567 (AF030027), EHV-8 strain wh (JQ343919), and EHV-9 strain P19 (AP010838), with each genome size as shown in Table 1. The determined genome sizes were consistent with the sequences; however, they are unlikely to accurately reflect the actual sizes. This discrepancy arises because each genome sequence contains numerous direct tandem repeats of short sequences, often arranged in complex configurations that include partial or dispersed repeats. In the case of αEHVs, these reiterations (repeats) frequently displayed variability in size, contributing to the heterogeneity observed in genome sizes. Moreover, the tandem repeats located at the genome termini exhibit a high GC content, similar to that observed in telomeric regions. CpG islands are defined by a high concentration of CG dinucleotides, and their relative scarcity is a well-documented evolutionary phenomenon. This scarcity results from the methylation of cytosine residues, which subsequently undergo spontaneous deamination to form thymine, leading to the formation a TG dinucleotide [30]. This mutation can become fixed in the genome through replication processes. Such mechanisms have been observed in herpesviruses during studies investigating the methylation patterns of latent genomes within proliferating cell populations [31]. A class of tandem repeats has been identified within the genomes of Marek's disease virus and human herpesviruses (HHVs) 6A, 6B, and 7 [31], which may facilitate the integration of the viral genome into the telomeric regions of host eukaryotic chromosomes [32,33]. However, it remains uncertain whether this mechanism is applicable to αEHVs.

Table 1. Genomic arrangement and size of αEHVs in subfamily *Alphaherpesvirinae*.

Name	Size (bp)					G + C (%)	Total Numbers of ORFs				
	Genome	U_L [a]	U_S [b]	IR [c]	TR [d]		Genome	U_L	U_S	IR	TR
EHV-1	150,224	112,935	11,861	12,714	12,714	56.67	80	62	10	4	4
EHV-4	145,597	112,452	12,789	10,179	10,179	50.46	79	62	10	4	3
EHV-8	149,332	113,341	89,403	11,969	11,969	54.36	80	62	10	4	4
EHV-9	148,371	112,681	11,998	11,846	11,846	56.10	80	62	10	4	4

[a], long unique region; [b], short unique region; [c], inverted repeat; [d], terminal inverted repeat.

Potential protein-coding regions within the EHV-1 genome were initially identified through the search for ATG-initiated open reading frames (ORFs), with a specific focus on evaluating codon usage preferences and the G + C bias at the third codon position. Additionally, general characteristics of the gene organization were taken into account, such as the limited overlap between ORFs and the positioning of potential polyadenylation signals (AATAAA, ATTAAA, and AGTAAA), which are transcribed as part of 3′-coterminal families. This information pertaining to the EHV-1 genome was subsequently utilized to identify EHV-4 genes in an earlier study.

The genome of these viruses comprises unique long (UL), unique short (US), inverted repeat (IR), and terminal inverted repeat (TR) regions. αEHVs are biologically categorized within the *Alphaherpesvirinae* subfamily, which also includes herpes simplex virus type 1 (HSV-1), varicella zoster virus (VZV), and the pseudorabies virus (PRV). Previous studies have detailed the DNA sequences, structural characteristics, and genome mapping of EHV-1 and EHV-4, highlighting their collinear arrangement similarities with HSV-1 and VZV [34,35]. EHV-8 (Figure 1) and EHV-9 (Figure S1) have been identified as possessing a genomic structure analogous to that of EHV-1. In the current study, the four EHV genomes analyzed comprised four distinct components, the dimensions of which are detailed in Table 1. All genomes exhibited highly collinear structures, characterized by a UL region flanked by a short IR linked to a US region, followed by a substantial TR. Notably, with the exception of a single copy of ORF67 in the EHV-4 genome, the other three genomes contained duplicated TRs and IRs for ORF64, ORF65, ORF66, and ORF67. ORF35 was expressed as a full-length mRNA from an upstream promoter, as well as a 3′-coterminal truncated mRNA initiated internally. The protein encoded by the smaller overlapping gene, designated ORF35.5, was identical to the carboxy-terminal portion of the protein encoded by ORF35.

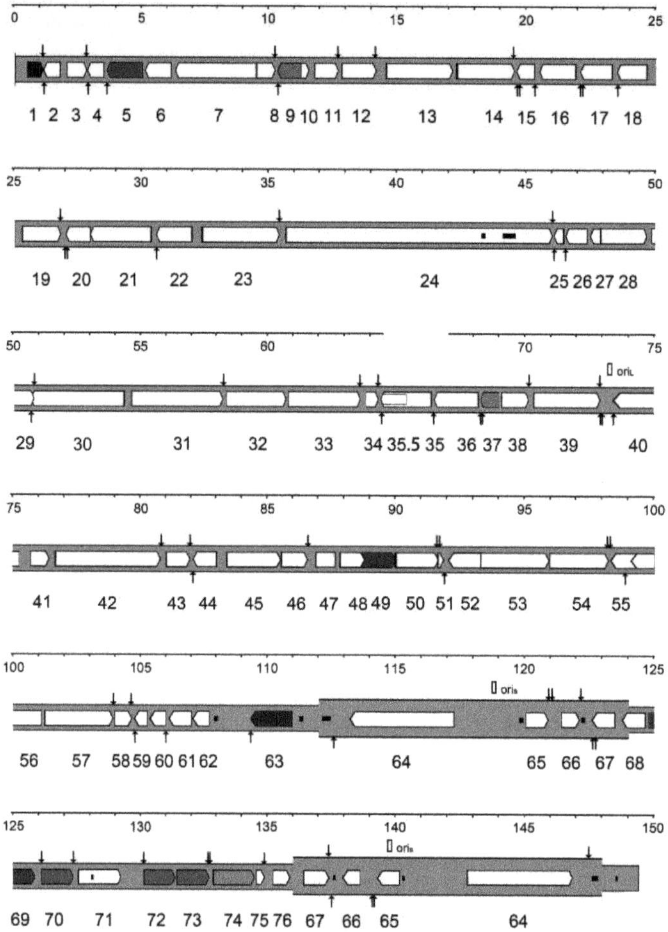

Figure 1. The genomic map of EHV-8. The terminal direct repeat (TR) is shown in a thicker format than the unique region (U). ORFs predicted to encode functional proteins are indicated by arrows (see the key below), with the nomenclature without the ORF prefix given below. The poly(A) sites are indicated by vertical arrows above and below the genome for ORFs oriented toward the right and left, respectively. Reiterated sequences are shown as small filled rectangles, and candidate origins of DNA replication (open squares) are indicated above the genome.

The nomenclature for the ORFs of EHV-8 is derived from that of EHV-1, EHV-4, and EHV-9, with ORFs sequentially numbered from 1 to 76. The ORFs of EHV-8 and EHV-9 are designed in accordance with their homologous counterparts in EHV-1 and EHV-4. In contrast, the nomenclature systems used for other alphaherpesviruses differ from that of the αEHVs, which typically exhibit a dense arrangement of ORFs. Several genomic regions are not anticipated to encode functional proteins; these regions are located at the left and right terminus of αEHVs genome, as well as between ORF62 and ORF63, ORF63 and ORF64, and ORF64 and ORF65. The predicted counts of functional protein-coding ORFs in the four genomes represent the most accurate estimates currently available.

3. Gene Function

The potential functions of each protein, as presented and proposed in Table S1, were elucidated through the findings of the current study by integrating known characteristics of the members in *Orthoherpesviridae*, with HSV-1, VZV, and PRV as example.

The current understanding of αEHV genomic function is limited. Existing researches predominantly concentrate on the gene function of EHV-1. EHV-1 utilizes both viral transcription regulatory proteins and host transcription factors to support its infection in vivo and in vitro. The transcriptional regulation of the EHV-1 genome is orchestrated in a temporally controlled cascade, categorized into immediate-early (IE), early (E), and late (L) phases during lytic infection [36]. The sole IE gene, transcribed first, encodes the regulatory protein IEP, which is critical for initiating subsequent viral gene expression. Early genes further prime the cellular environment for viral replication, while late genes, dependent on viral DNA synthesis, encode structural components like glycoproteins (e.g., gB and gC). For example, the EHV-1 IE protein homolog (similar to HSV-1 ICP4) activates E genes, including those involved in DNA replication [37]. The expression of EHV-1 genes over time is intricately regulated by six distinct molecules: the sole IE protein, four early regulatory proteins (EICP0, EICP22, EICP27, and IR2P), and a late ETIF protein [36]. The IE protein enhances the activity of EHV-1 promoters by binding to the consensus sequence 5′-ATCGT-3′, which is found within these promoters. This binding recruits additional TATA-binding proteins (TBP) and transcription factor IIB (TFIIB) to help establish the transcription initiation complex. Conversely, IR2P serves as a negative regulator, diminishing promoter *cis*-activity by depleting the availability of essential cellular transcription factors [38,39], which leads to a significant downregulation in both gene expression and viral titers. The tegument protein ETIF specifically and exclusively activates the IE gene promoter through interactions with transcription factor Oct-1, which targets octamer sequences [40,41]. EICP0 has the capability to independently activate all types of EHV-1 gene promoters by interacting with TBP and TFIIB as well as the simplest promoter only containing TATA site and transcription start site [42,43]. Meanwhile, EICP22 (IR4) [44–46] and EICP27 [47,48] each have a minimal trans-activation effect on promoters. However, when they work together, they enhance the activation of the promoter of the IE gene, acting as co-factors that boost IEP activity. EICP22 increases the in vitro host spectrum of EHV-1 and plays an important role in pathogenesis within the murine model. EICP27 synergizes with EICP0 to activate both early and leaky late promoters, and it also directly interacts with TBP. However, EICP22 and EICP0 do not function collaboratively to influence any EHV-1 promoters. Additionally, UL4 interacts with cellular transcription factors, such as the TBP to modulate viral gene expression [49].

The viral DNA polymerase (ORF30) not only facilitates DNA replication but may also engage with cellular RNA polymerase II to enhance transcription processes to *trans*-activate L genes. The polymorphisms in ORF30, being regarded as neuropathogenic markers (e.g., variants harboring A2254/N752 and C2254/H752), have been associated with variations in transcriptional efficiency and clinical outcomes, indicating genotype-specific regulatory differences [50]. The genotype of EHV-1 DNA polymerase (ORF30) [51], specifically variants D752, N752, and H752, correlates with diverse pathogenic profiles [9]; yet, the exact molecular mechanisms remain elusive. Recent genomic studies of EHV-1 outbreaks in Sweden (2012–2021) identified 14 geno-variants, along with new mutations in ORF11 and ORF34 that influence viral fitness [50]. Synonymous mutations in ORF11 and non-synonymous alterations in ORF34 (mutable genomic locus) may modify the secondary structures mRNA or influence protein interactions, impacting transcriptional efficiency [50]. The regulation of EHV-1 transcription is a complex process affected by viral genetic variability, epigenetic alterations, and dynamic protein interactions. Comparative studies with other herpesviruses (e.g., HSV-1) create opportunities to explore unresolved molecular mechanisms. Future inquiries should focus on in vivo models to confirm transcriptional regulators as potential therapeutic targets, especially concerning neuropathogenic strains and approaches to control latency.

Numerous functional proteins require further investigation, identification, and validation. Certain glycoproteins (e.g., gD, gB, gE, and gC) have been shown to facilitate viral entry into host cells, mediate cell-to-cell transportation, elicit protective immune responses, and modulate immune response and are associated with virulence [52–60]. The proteins UL43, UL49.5, and UL56 have been shown to downregulate MHC I levels via inhibiting TAP binding [61–65]. EHV-1 UL45 (ORF15) is identified as a virulence factor contributing to neuropathogenesis and influencing viral replication [66,67]. EHV-1 ORF1 and ORF2 exhibit a close relation with virulence during infection in horses [68,69]. US2 (ORF68) modulates virus entry and cell-to-cell spread and relates to virulence [70,71].

EHV-1 expresses two non-coding RNA: latency-associated transcripts (LATs) and IR3. Latency is characterized by the suppression of lytic gene transcription and maintenance of the viral genome via LATs [72,73]. In murine models, EHV-1 reactivation from latency following corticosteroid administration correlated with transient viremia and respiratory shedding, underscoring the reversibility of latency [74]. LATs play a crucial role in establishing latent EHV infections within the host. IR3 suppresses the expression of IE gene, thereby influencing the biological characteristics and virulence of EHV-1. The IR3 transcript of equine herpesvirus-1 (EHV-1) contains 117 nucleotides that are antisense to the IE mRNA, indicating a potential regulatory function [75–77].

It is plausible that some of these ORFs do not encode functional proteins, while others may have been overlooked. These ORFs are small, spliced, weakly conserved, or entirely non-conserved and use non-classical initiation codons (i.e., non-ATG), a phenomenon proposed in human herpesvirus (e.g., HSV-2 UL16) [78]. Translational initiation may take place from a codon that is not ATG, which has been demonstrated to serve as an initiation codon in eukaryotic systems [79,80]. Furthermore, although not yet identified, overlapping ORFs may exist within αEHVs genomes. Notably, two ORFs have been identified within the intron located in the 5′ untranslated region (UTR) of the IE gene [81].

4. Genetic Similarity

EHV-1 exhibits the largest genome among the four viruses. Furthermore, a homology analysis of the nucleotide and amino acid sequences was conducted for each gene common to the four viruses (Table S2), as any genetic diversity is likely to result in distinct characteristics in the interaction of αEHVs. The pairwise alignment between homologous ORFs and their products (amino acid sequence), annotated in EHV-1, EHV-4, EHV-8, and EHV-9, were performed using the Clustal W algorithm of the MegAlign platform with DNAstar (version 7.1) software.

4.1. Conserved ORFs in Orthoherpesviridae

Members of the *Orthoherpesviridae*, within the order *Herpesvirales*, possess 44 shared genes (referred to as core ORFs, Table S1), which are seemingly inherited from a common ancestor [82]. This genetic composition highlights a greater divergence within the *Herpesviridae* family compared to the *Alloherpesviridae* family, which contains only 12 core genes [31]. These genes encompass the majority of functional categories outlined in Table S1, including those associated with nucleotide metabolism, DNA replication, capsid structure, and virion morphogenesis.

The ORF44/47 encoding the DNA packaging terminase subunit 1 exhibits a high degree of conservation across members of the order *Herpesvirales* [83]. Conversely, the ORF encoding the DNA packaging terminase subunit 2 (ORF32) is conserved within the *Orthoherpesviridae* but demonstrates only limited similarity to the core gene (ORF47) in cyprinid herpesvirus, with no analogous sequences identified in other alloherpesviruses [31]. Notably, the conserved region encompasses a cysteine-rich motif, which is ubiquitous among *Orthoherpesviridae* members and may function as a metal ion-binding domain [84–86].

4.2. Conserved Genes in Subfamily *Alphaherpesvirinae*

In the *Alphaherpesvirinae* subfamily, a total of 24 conserved genes were identified except for 44 core genes. Of these, 44 core genes were inherited from the common ancestor shared by the α-, β-, and γ-subfamilies of *Orthoherpesviridae*, while these 24 genes were inherited from an ancestor specific to the *Alphaherpesvirinae* subfamily [82], as indicated in Table S1 using orthologs. This phylogenetic classification suggests that specific core or orthologous genes may have been lost in certain lineages. In the case of αEHVs, 11 viral glycoproteins [87], including gB, gC, gD, gE, gG, gH, gI, gK, gL, gM, and gN, are conserved when compared to other alphaherpesviruses. The glycoprotein gp2 (gJ) is encoded by ORF71 exclusively in αEHVs and is located within the US genomic segment [88–90]. EHV-1 strain Kentucky A (KyA) harbors a 1242 bp deletion and encodes a truncated gp2 of only 383 amino acids [89,90], in contrast to 797 and 791 amino acids in the pathogenic strains Ab4 and RacL11, respectively. As regulatory proteins, ORF64 (IEP), truncated ORF64 (IR2P), ORF12 (ETIF), and ORF5 (EICP27) are conserved in *Alphaherpesvirinae*. The remaining eight genes—ORF2, ORF3, ORF4, ORF34, ORF63 (EICP0), 65 (EICP22), ORF67 and ORF76—are not conserved in this subfamily. Of the 76 ORFs, there is no knowledge to date about potential functions associated with ORF4, ORF58, ORF60, ORF66, and ORF75.

4.3. Orthologs in HSV-1 and VZV

The αEHVs exhibit 71 and 70 orthologs in common with HSV-1 and VZV, respectively (Table S1). The degree of genome collinearity between αEHVs and either HSV-1 or VZV is comparable to that observed among αEHVs themselves, characterized by extensive blocks that are rearranged in terms of order and orientation relative to one another. These findings underscore the close evolutionary relationship among αEHVs. It is important to note that while αEHVs are considered to demonstrate a pattern of recent coevolution with their hosts, this does not necessarily imply cospeciating with their hosts, nor with HSV-1 and VZV [34,35]. Therefore, within the *Alphaherpesvirinae* subfamily, αEHVs, HSV-1, and VZV exhibit a closer phylogenetic relationship to one another than to their respective host species, suggesting that one of these lineages likely evolved through an interspecies transfer event. Future and ongoing studies may be informed and guided by the relevant information associated with orthologs from HSV-1 and VZV.

4.4. Fragmented ORF

EHV-1 and EHV-4 have been documented to possess a single fragmented open reading frame (ORF 44/47), while the EHV-8 and EHV-9 equivalents have been annotated in GenBank, albeit containing an intergenic region that is highly conserved among αEHVs, which are homologous to the two exons of HSV-1 UL15, which is expressed as a spliced mRNA.

4.5. UL24 Protein Family in Orthoherpesviridae

The UL24 protein family is conserved across all three subfamilies of the *Orthoherpesviridae* [91]. This conservation suggests the existence of the same ancestor gene that dates back at least 180 million years, prior to the emergence of the three subfamilies [92,93]. The UL24 family encodes a variety of potential PD-(D/E)XK endonucleases, characterized by the presence of signature motifs II, III, and IV, which are part of a broad and diverse superfamily of restriction endonucleases and recombinases [94,95]. The UL24 protein can inhibit the innate immune response of the host through its interaction with various immune signaling pathways (e.g., RIG-I, cGAS-STING) involved in a series of cellular factors, including p65, p50, IRF7, ISG20, IL-8, OASL, IL-β, and ZCCHZ3, during the course of a viral infection. This multifaceted role not only allows the virus to evade the initial defenses of the host immune system but also contributes significantly to the ability of herpesvirus to replicate and become pathogenic in the later stages of the infection [91,95]. UL24 is crucial for the herpesvirus lifecycle, enhancing both its proliferation and its capacity to cause disease in the category of *Orthoherpesviridae*.

In HHVs, the UL24 protein induces cell cycle arrest by inactivating the cyclin B/cdc2 complex and may also function as a minor tegument component that is weakly associated with capsids [96]. The pervasive presence of the UL24 the *Orthoherpesviridae* indicates its essential function in the viral life cycle, particularly in processes such as membrane fusion and replication [95]. This role is evidenced in tissue culture by the dispersal and redistribution of the nucleolar protein (NPM1) [96–98]. Currently, there are only two documented studies concerning UL24 in EHV-1. One study, utilizing a mouse model, identified UL24 as a determinant of neuropathogenicity, while another study provided sequence data for its homolog from Brazilian isolates of EHV-1 [99,100]. In equid herpesviruses, the biological function of the UL24 protein warrants further investigation, and elucidating the associated molecular mechanisms may be facilitated by leveraging research conducted on other herpesviruses within the same family. EHV-1 UL24/TK (head to head located genome) deletion may be the prospective target to break latent infection and pave the novel way to develop vaccine and antiviral reagents aiming at the infection of αEHVs [101].

5. Problems and Possible Solutions

Currently, there are no immortalized cell lines derived from the respiratory and nervous systems of equids, which poses significant challenges to understanding the molecular mechanisms underlying the pathogenesis of αEHVs in vivo. In the context of αEHV research, commonly used cell lines such as RK13, BHK-21, CHO, and MDBK can support the replication of EHV-1, EHV-3, EHV-8, and EHV-9 in vitro; however, these cell lines exhibit considerable genetic diversity when compared to equine cells. Fetal horse kidney cells (FHK) and equid dermis cells (ED) are utilized in certain studies related to EHV-4, albeit within a limited number of passages.

The transformation of FHK cells with primate genes, including large T antigen of SV40 (LTA), results in their immortalization, allowing them to be employed for the isolation and propagation of equine herpesvirus [102–104]. The fields of genomics, transcriptomics, proteomics, and metabolomics in equids are currently underdeveloped, resulting in a lack of comprehensive information regarding gene function. This shortcoming poses a significant barrier to advancing our understanding of the molecular interactions between EHVs and horses. Consequently, the equine telomere reverse transcriptase (eTERT) gene may be viewed as a potential catalyst for the development of immortalized equine cell lines and oncogenic genes associated with equine adenovirus, similar to the established immortalized human cell lines. The establishment of such cell lines could greatly enhance research efforts focused not only on EHVs but also on other equine pathogens.

6. Conclusions and Prospects

This review provides a systemic summary of the genomic information of αEHVs, focusing on their structural and functional characteristics, highlighting the identity and diversity among homologs through comparative alignment. This review represents the initial investigation of a collection of closely related αEHV genomes and aims to enhance the understanding of αEHVs within the *Orthoherpesviridae* family to match the depth of knowledge available for the far more thoroughly researched Herpesviridae family. Similar to what is becoming evident with human herpesviruses (e.g., HSV-1 and VZV), the genomic characteristics of αEHVs will provide a foundation for forthcoming research on the pathogenesis and immune evasion mechanisms of αEHVs and for efforts focused on vaccine and therapeutics, diagnostics, and epidemiological advancements.

In the context of current biological research, initial investigations might prioritize targeting αEHVs genes labeled as "unknown" and "possibly" in terms of function, as indicated in Table S1. The observed discrepancies between homologs may facilitate the establishment of differential diagnoses. Investigating, detecting, identifying, and confirming novel transcription units within the αEHV genome represent significant research avenues, particularly because certain AATAAA sequences have not been annotated as poly(A) signal sites in GenBank data. These AATAAA motifs likely suggest the presence of previously uncharacterized transcripts. Also, future research should aim to identify and confirm non-essential genes involved in viral replication in vitro. By targeting these regions for genetic editing, the genome of αEHVs could be optimized as a live vector for the delivery of heterologous antigens, thereby enhancing strategies for the prevention and treatment of infectious diseases in equids. In summary, this review aims to advance research on αEHVs within the current context.

Supplementary Materials: The following supporting information can be downloaded at: https://www.mdpi.com/article/10.3390/vetsci12030228/s1, Figure S1: The genomic map of EHV-9. The terminal direct repeat (TR) is shown in a thicker format than the unique region (U). ORFs predicted to encode functional proteins are indicated by arrows (see the key below), with the nomenclature without the ORF prefix given below. The poly(A) sites are indicated by vertical arrows above and below the genome for ORFs oriented toward the right and left, respectively. Reiterated sequences are shown as small filled rectangles, and candidate origins of DNA replication (open squares) are indicated above the genome; Table S1: Features of functional protein-coding regions of EHVs in *Alphaherpesvirinae*; Table S2: Alignment of protein-coding regions of EHVs in *Alphaherpesvirinae* (%).

Author Contributions: Writing—original draft preparation, D.L. and X.Z.; review and editing, D.L. and X.W. All authors have read and agreed to the published version of the manuscript.

Funding: This research was funded by Heilongjiang Provincial Natural Science Foundation of China (grant number LH2022C109) and General Program of National Natural Science Foundation of China (grant number 32272973).

Institutional Review Board Statement: Not applicable.

Informed Consent Statement: Not applicable.

Data Availability Statement: The original contributions presented in this study are included in this article and supplementary material. Further inquiries can be directed to the corresponding authors.

Conflicts of Interest: The authors declare no conflicts of interest.

References

1. Gatherer, D.; Depledge, D.P.; Hartley, C.A.; Szpara, M.L.; Vaz, P.K.; Benko, M.; Brandt, C.R.; Bryant, N.A.; Dastjerdi, A.; Doszpoly, A.; et al. ICTV Virus Taxonomy Profile: Herpesviridae 2021. *J. Gen. Virol.* **2021**, *102*, 10. [CrossRef] [PubMed]

2. Pellett, P.E.; Davison, A.J.; Eberle, R.; Ehlers, B.; Hayward, G.S.; Lacoste, V.; Minson, A.C.; Nicholas, J.; Roizman, B.; Studdert, M.J.; et al. Order-Herpesvirales. In *Virus Taxonomy: Classification and Nomenclature of Viruses: Ninth Report of the International Committee on Taxonomy of Viruses*; King, A.M.Q., Adams, M.J., Carstens, E.B., Lefkowitz, E.J., Eds.; Elsevier: San Diego, CA, USA, 2012; pp. 99–107.
3. Baumann, R.P.; Sullivan, D.C.; Staczek, J.; O'Callaghan, D.J. Genetic relatedness and colinearity of genomes of equine herpesvirus types 1 and 3. *J. Virol.* **1986**, *57*, 816–825. [CrossRef] [PubMed]
4. Davison, A.J.; Eberle, R.; Ehlers, B.; Hayward, G.S.; McGeoch, D.J.; Minson, A.C.; Pellett, P.E.; Roizman, B.; Studdert, M.J.; Thiry, E. The order Herpesvirales. *Arch. Virol.* **2009**, *154*, 171–177. [CrossRef] [PubMed]
5. Khusro, A.; Aarti, C.; Rivas-Caceres, R.R.; Barbabosa-Pliego, A. Equine Herpesvirus-I Infection in Horses: Recent Updates on its Pathogenicity, Vaccination, and Preventive Management Strategies. *J. Equine Vet. Sci.* **2020**, *87*, 102923. [CrossRef]
6. Ivens, P.; Rendle, D.; Kydd, J.; Crabtree, J.; Moore, S.; Neal, H.; Knapp, S.; Bryant, N.; Newton, J.R. Equine herpesviruses: A roundtable discussion. *UK-Vet. Equine* **2019**, *3* (Suppl. 4), 1–12. [CrossRef]
7. Patel, J.R.; Heldens, J. Equine herpesviruses 1 (EHV-1) and 4 (EHV-4)—Epidemiology, disease and immunoprophylaxis: A brief review. *Vet. J.* **2005**, *170*, 14–23. [CrossRef]
8. Afify, A.F.; Hassanien, R.T.; El Naggar, R.F.; Rohaim, M.A.; Munir, M. Unmasking the ongoing challenge of equid herpesvirus-1 (EHV-1): A comprehensive review. *Microb. Pathog.* **2024**, *193*, 106755. [CrossRef]
9. Pusterla, N.; Barnum, S.; Lawton, K.; Wademan, C.; Corbin, R.; Hodzic, E. Investigation of the EHV-1 Genotype (N 752, D 752, and H 752) in Swabs Collected from Equids with Respiratory and Neurological Disease and Abortion from the United States (2019–2022). *J. Equine Vet. Sci.* **2023**, *123*, 104244. [CrossRef]
10. O'Callaghan, D.J.; Osterrieder, N. Herpesviruses of Horses. In *Encyclopedia of Virology*, 3rd ed.; Mahy, B.W.J., van Regenmortel, M., Eds.; Academic Press: Oxford, UK, 2008; pp. 411–420.
11. Pavulraj, S.; Eschke, K.; Theisen, J.; Westhoff, S.; Reimers, G.; Andreotti, S.; Osterrieder, N.; Azab, W. Equine Herpesvirus Type 4 (EHV-4) Outbreak in Germany: Virological, Serological, and Molecular Investigations. *Pathogens* **2021**, *10*, 810. [CrossRef]
12. Lefkowitz, E.J.; Dempsey, D.M.; Hendrickson, R.C.; Orton, R.J.; Siddell, S.G.; Smith, D.B. Virus taxonomy: The database of the International Committee on Taxonomy of Viruses (ICTV). *Nucleic Acids Res.* **2018**, *46*, D708–D717. [CrossRef]
13. Thormann, N.; Van de Walle, G.R.; Azab, W.; Osterrieder, N. The role of secreted glycoprotein G of equine herpesvirus type 1 and type 4 (EHV-1 and EHV-4) in immune modulation and virulence. *Virus Res.* **2012**, *169*, 203–211. [CrossRef] [PubMed]
14. Browning, G.F.; Ficorilli, N.; Studdert, M.J. Asinine herpesvirus genomes: Comparison with those of the equine herpesviruses. *Arch. Virol.* **1988**, *101*, 183–190. [CrossRef] [PubMed]
15. Wang, T.T.; Xi, C.K.; Yu, Y.; Liu, W.Q.; Akhtar, M.F.; Li, Y.B.; Wang, C.F.; Li, L.L. Characteristics and epidemiological investigation of equid herpesvirus 8 in donkeys in Shandong, China. *Arch. Virol.* **2023**, *168*, 3. [CrossRef]
16. Wang, T.T.; Hu, L.Y.; Wang, Y.H.; Liu, W.Q.; Liu, G.Q.; Zhu, M.X.; Zhang, W.; Wang, C.F.; Ren, H.Y.; Li, L.L. Identification of equine herpesvirus 8 in donkey abortion: A case report. *Virol. J.* **2022**, *19*, 1. [CrossRef] [PubMed]
17. Wang, T.T.; Hu, L.Y.; Liu, M.Y.; Wang, T.J.; Hu, X.Y.; Li, Y.; Liu, W.Q.; Li, Y.B.; Wang, Y.H.; Ren, H.Y.; et al. The Emergence of Viral Encephalitis in Donkeys by Equid Herpesvirus 8 in China. *Front. Microbiol.* **2022**, *13*, 840754. [CrossRef]
18. Garvey, M.; Suarez, N.M.; Kerr, K.; Hector, R.; Moloney-Quinn, L.; Arkins, S.; Davison, A.J.; Cullinane, A. Equid herpesvirus 8: Complete genome sequence and association with abortion in mares. *PLoS ONE* **2018**, *13*, e0192301. [CrossRef]
19. Liu, C.; Guo, W.; Lu, G.; Xiang, W.; Wang, X. Complete genomic sequence of an equine herpesvirus type 8 Wh strain isolated from China. *J. Virol.* **2012**, *86*, 5407. [CrossRef]
20. Crabb, B.S.; Allen, G.P.; Studdert, M.J. Characterization of the major glycoproteins of equine herpesviruses 4 and 1 and asinine herpesvirus 3 using monoclonal antibodies. *J. Gen. Virol.* **1991**, *72 Pt 9*, 2075–2082. [CrossRef]
21. Crabb, B.S.; Studdert, M.J. Comparative studies of the proteins of equine herpesviruses 4 and 1 and asinine herpesvirus 3: Antibody response of the natural hosts. *J. Gen. Virol.* **1990**, *71 Pt 9*, 2033–2041. [CrossRef]
22. Taniguchi, A.; Fukushi, H.; Matsumura, T.; Yanai, T.; Masegi, T.; Hirai, K. Pathogenicity of a new neurotropic equine herpesvirus 9 (gazelle herpesvirus 1) in horses. *J. Vet. Med. Sci.* **2000**, *62*, 215–218. [CrossRef]
23. Paillot, R.; Sharp, E.; Case, R.; Nugent, J. Herpes Virus Infection in Equid Species. In *Herpesviridae: Viral Structure, Life Cycle and Infections*; Nova Biomedical: Runcorn, UK, 2009; pp. 17–85.
24. Taniguchi, A.; Fukushi, H.; Yanai, T.; Masegi, T.; Yamaguchi, T.; Hirai, K. Equine herpesvirus 9 induced lethal encephalomyelitis in experimentally infected goats. *Arch. Virol.* **2000**, *145*, 2619–2627. [CrossRef] [PubMed]
25. El-Habashi, N.; Murakami, M.; El-Nahass, E.; Hibi, D.; Sakai, H.; Fukushi, H.; Sasseville, V.; Yanai, T. Study on the infectivity of equine herpesvirus 9 (EHV-9) by different routes of inoculation in hamsters. *Vet. Pathol.* **2011**, *48*, 558–564. [CrossRef] [PubMed]
26. Schrenzel, M.D.; Tucker, T.A.; Donovan, T.A.; Busch, M.D.; Wise, A.G.; Maes, R.K.; Kiupel, M. New hosts for equine herpesvirus 9. *Emerg. Infect. Dis.* **2008**, *14*, 1616–1619. [CrossRef] [PubMed]
27. Yanai, T.; Tujioka, S.; Sakai, H.; Fukushi, H.; Hirai, K.; Masegi, T. Experimental infection with equine herpesvirus 9 (EHV-9) in cats. *J. Comp. Pathol.* **2003**, *128*, 113–118. [CrossRef] [PubMed]

28. Yanai, T.; Fujishima, N.; Fukushi, H.; Hirata, A.; Sakai, H.; Masegi, T. Experimental infection of equine herpesvirus 9 in dogs. *Vet. Pathol.* **2003**, *40*, 263–267. [CrossRef]
29. Kodama, A.; Yanai, T.; Kubo, M.; El-Habashi, N.; Kasem, S.; Sakai, H.; Masegi, T.; Fukushi, H.; Kuraishi, T.; Yoneda, M.; et al. Cynomolgus monkeys (Macaca fascicularis) may not become infected with equine herpesvirus 9. *J. Med. Primatol.* **2011**, *40*, 18–20. [CrossRef]
30. Honess, R.W.; Gompels, U.A.; Barrell, B.G.; Craxton, M.; Cameron, K.R.; Staden, R.; Chang, Y.N.; Hayward, G.S. Deviations from expected frequencies of CpG dinucleotides in herpesvirus DNAs may be diagnostic of differences in the states of their latent genomes. *J. Gen. Virol.* **1989**, *70 Pt 4*, 837–855. [CrossRef]
31. Davison, A.J.; Kurobe, T.; Gatherer, D.; Cunningham, C.; Korf, I.; Fukuda, H.; Hedrick, R.P.; Waltzek, T.B. Comparative genomics of carp herpesviruses. *J. Virol.* **2013**, *87*, 2908–2922. [CrossRef]
32. Arbuckle, J.H.; Medveczky, M.M.; Luka, J.; Hadley, S.H.; Luegmayr, A.; Ablashi, D.; Lund, T.C.; Tolar, J.; De Meirleir, K.; Montoya, J.G.; et al. The latent human herpesvirus-6A genome specifically integrates in telomeres of human chromosomes in vivo and in vitro. *Proc. Natl. Acad. Sci. USA* **2010**, *107*, 5563–5568. [CrossRef]
33. Kaufer, B.B.; Jarosinski, K.W.; Osterrieder, N. Herpesvirus telomeric repeats facilitate genomic integration into host telomeres and mobilization of viral DNA during reactivation. *J. Exp. Med.* **2011**, *208*, 605–615. [CrossRef]
34. Telford, E.A.; Watson, M.S.; Perry, J.; Cullinane, A.A.; Davison, A.J. The DNA sequence of equine herpesvirus-4. *J. Gen. Virol.* **1998**, *79 Pt 5*, 1197–1203. [CrossRef]
35. Telford, E.A.; Watson, M.S.; McBride, K.; Davison, A.J. The DNA sequence of equine herpesvirus-1. *Virology* **1992**, *189*, 304–316. [CrossRef] [PubMed]
36. Ma, Y.; Liu, D.; Gao, J.; Wang, X. Similar regulation of two distinct UL24 promoters by regulatory proteins of equine herpesvirus type 1 (EHV-1). *FEBS Lett.* **2015**, *589*, 1467–1475. [CrossRef]
37. Okada, A.; Suganuma, S.; Badr, Y.; Omatsu, T.; Mizutani, T.; Ohya, K.; Fukushi, H. Decreased expression of the immediate early protein, ICP4, by deletion of the tegument protein VP22 of equine herpesvirus type 1. *J. Vet. Med. Sci.* **2018**, *80*, 311–315. [CrossRef]
38. Kim, S.K.; Kim, S.; Dai, G.; Zhang, Y.; Ahn, B.C.; O'Callaghan, D.J. Identification of functional domains of the IR2 protein of equine herpesvirus 1 required for inhibition of viral gene expression and replication. *Virology* **2011**, *417*, 430–442. [CrossRef] [PubMed]
39. Kim, S.K.; Ahn, B.C.; Albrecht, R.A.; O'Callaghan, D.J. The unique IR2 protein of equine herpesvirus 1 negatively regulates viral gene expression. *J. Virol.* **2006**, *80*, 5041–5049. [CrossRef] [PubMed]
40. Hofmann-Sieber, H.; Wild, J.; Fiedler, N.; Tischer, K.; von Einem, J.; Osterrieder, N.; Hofmann, H.; Koestler, J.; Wagner, R. Impact of ETIF Deletion on Safety and Immunogenicity of Equine Herpesvirus Type 1-Vectored Vaccines. *J. Virol.* **2010**, *84*, 11602–11613. [CrossRef]
41. von Einem, J.; Schumacher, D.; O'Callaghan, D.J.; Osterrieder, N. The alpha t-TIF (VP16) homologue (ETIF) of equine herpesvirus 1 is essential for secondary envelopment and virus egress. *J. Virol.* **2006**, *80*, 2609–2620. [CrossRef]
42. Kim, S.K.; Albrecht, R.A.; O'Callaghan, D.J. A negative regulatory element (base pairs -204 to -177) of the EICP0 promoter of equine herpesvirus 1 abolishes the EICP0 protein's trans-activation of its own promoter. *J. Virol.* **2004**, *78*, 11696–11706. [CrossRef]
43. Kim, S.K.; Jang, H.K.; Albrecht, R.A.; Derbigny, W.A.; Zhang, Y.; O'Callaghan, D.J. Interaction of the equine herpesvirus 1 EICP0 protein with the immediate-early (IE) protein, TFIIB, and TBP may mediate the antagonism between the IE and EICP0 proteins. *J. Virol.* **2003**, *77*, 2675–2685. [CrossRef]
44. Derbigny, W.A.; Kim, S.K.; Jang, H.K.; O'Callaghan, D.J. EHV-1 EICP22 protein sequences that mediate its physical interaction with the immediate-early protein are not sufficient to enhance the trans-activation activity of the 1E protein. *Virus Res.* **2002**, *84*, 1–15. [CrossRef] [PubMed]
45. Derbigny, W.A.; Kim, S.K.; Caughman, G.B.; O'Callaghan, D.J. The EICP22 protein of equine herpesvirus 1 physically interacts with the immediate-early protein and with itself to form dimers and higher-order complexes. *J. Virol.* **2000**, *74*, 1425–1435. [CrossRef] [PubMed]
46. Breitenbach, J.E.; Ebner, P.D.; O'Callaghan, D.J. The IR4 auxiliary regulatory protein expands the in vitro host range of equine herpesvirus 1 and is essential for pathogenesis in the murine model. *Virology* **2009**, *383*, 188–194. [CrossRef]
47. Albrecht, R.A.; Kim, S.K.; O'Callaghan, D.J. The EICP27 protein of equine herpesvirus 1 is recruited to viral promoters by its interaction with the immediate-early protein. *Virology* **2005**, *333*, 74–87. [CrossRef] [PubMed]
48. Albrecht, R.A.; Kim, S.K.; Zhang, Y.; Zhao, Y.; O'Callaghan, D.J. The equine herpesvirus 1 EICP27 protein enhances gene expression via an interaction with TATA box-binding protein. *Virology* **2004**, *324*, 311–326. [CrossRef]
49. Zhang, Y.F.; Charvat, R.A.; Kim, S.K.; O'Callaghan, D.J. The EHV-1 UL4 protein that tempers viral gene expression interacts with cellular transcription factors. *Virology* **2014**, *449*, 25–34. [CrossRef]
50. Öhrmalm, J.; Cholleti, H.; Theelke, A.K.; Berg, M.; Gröndahl, G. Divergent strains of EHV-1 in Swedish outbreaks during 2012 to 2021. *BMC Vet. Res.* **2024**, *20*, 1. [CrossRef]

51. Sutton, G.; Thieulent, C.; Fortier, C.; Hue, E.S.; Marcillaud-Pitel, C.; Pléau, A.; Deslis, A.; Guitton, E.; Paillot, R.; Pronost, S. Identification of a New Equid Herpesvirus 1 DNA Polymerase (ORF30) Genotype with the Isolation of a C/H Strain in French Horses Showing no Major Impact on the Strain Behaviour. *Viruses* **2020**, *12*, 1160. [CrossRef]
52. Fuentealba, N.A.; Bravi, M.E.; Panei, C.J.; Zanuzzi, C.N.; Sguazza, G.H.; Corva, S.G.; Pecoraro, M.R.; Galosi, C.M. Production of recombinant glycoprotein C from equid alphaherpesvirus 1 and evaluation of the immune response induced in mice. *Analecta Vet.* **2021**, *41*, 1–18.
53. Rodrigues, P.R.C.; Cunha, R.C.; Santos, F.D.S.; Goncalves, V.S.; Albuquerque, P.M.M.; Santos, A.G.; Lima, M.; Leite, F.P.L. Expression and characterization of equid herpesvirus 1 glycoprotein D in Pichia pastoris. *Arq. Bras. Med. Vet. Zootec.* **2020**, *72*, 703–710. [CrossRef]
54. Spiesschaert, B.; Stephanowitz, H.; Krause, E.; Osterrieder, N.; Azab, W. Glycoprotein B of equine herpesvirus type 1 has two recognition sites for subtilisin-like proteases that are cleaved by furin. *J. Gen. Virol.* **2016**, *97*, 1218–1228. [CrossRef] [PubMed]
55. Andoh, K.; Takasugi, M.; Mahmoud, H.Y.; Hattori, S.; Terada, Y.; Noguchi, K.; Shimoda, H.; Bannai, H.; Tsujimura, K.; Matsumura, T.; et al. Identification of a major immunogenic region of equine herpesvirus-1 glycoprotein E and its application to enzyme-linked immunosorbent assay. *Vet. Microbiol.* **2013**, *164*, 18–26. [CrossRef] [PubMed]
56. Spiesschaert, B.; Osterrieder, N.; Azab, W. Comparative analysis of glycoprotein B (gB) of equine herpesvirus type 1 and type 4 (EHV-1 and EHV-4) in cellular tropism and cell-to-cell transmission. *Viruses* **2015**, *7*, 522–542. [CrossRef] [PubMed]
57. Azab, W.; Zajic, L.; Osterrieder, N. The role of glycoprotein H of equine herpesviruses 1 and 4 (EHV-1 and EHV-4) in cellular host range and integrin binding. *Vet. Res.* **2012**, *43*, 61. [CrossRef]
58. Kydd, J.H.; Hannant, D.; Robinson, R.S.; Bryant, N.; Osterrieder, N. Vaccination of foals with a modified live, equid herpesvirus-1 gM deletion mutant (RacHDeltagM) confers partial protection against infection. *Vaccine* **2020**, *38*, 388–398. [CrossRef]
59. Azab, W.; El-Sheikh, A. The role of equine herpesvirus type 4 glycoprotein k in virus replication. *Viruses* **2012**, *4*, 1258–1263. [CrossRef]
60. Neubauer, A.; Osterrieder, N. Equine herpesvirus type 1 (EHV-1) glycoprotein K is required for efficient cell-to-cell spread and virus egress. *Virology* **2004**, *329*, 18–32. [CrossRef]
61. Huang, T.; Ma, G.G.; Osterrieder, N. Equine Herpesvirus 1 Multiply Inserted Transmembrane Protein pUL43 Cooperates with pUL56 in Downregulation of Cell Surface Major Histocompatibility Complex Class I. *J. Virol.* **2015**, *89*, 6251–6263. [CrossRef]
62. Said, A.; Azab, W.; Damiani, A.; Osterrieder, N. Equine herpesvirus type 4 UL56 and UL49.5 proteins downregulate cell surface major histocompatibility complex class I expression independently of each other. *J. Virol.* **2012**, *86*, 8059–8071. [CrossRef]
63. Verweij, M.C.; Lipinska, A.D.; Koppers-Lalic, D.; Van Leeuwen, W.F.; Cohen, J.I.; Kinchington, P.R.; Messaoudi, I.; Bienkowska-Szewczyk, K.; Ressing, M.E.; Rijsewijk, F.A.M.; et al. The Capacity of UL49.5 Proteins To Inhibit TAP Is Widely Distributed among Members of the Genus Varicellovirus. *J. Virol.* **2011**, *85*, 2351–2363. [CrossRef]
64. Verweij, M.C.; Lipinska, A.D.; Koppers-Lalic, D.; Quinten, E.; Funke, J.; van Leeuwen, H.C.; Bienkowska-Szewczyk, K.; Koch, J.; Ressing, M.E.; Wiertz, E.J.H.J. Structural and functional analysis of the TAP-inhibiting UL49.5 proteins of varicelloviruses. *Mol. Immunol.* **2011**, *48*, 2038–2051. [CrossRef] [PubMed]
65. Koppers-Lalic, D.; Verweij, M.C.; Lipinska, A.D.; Wang, Y.; Quinten, E.; Reits, E.A.; Koch, J.; Loch, S.; Marcondes Rezende, M.; Daus, F.; et al. Varicellovirus UL49.5 proteins differentially affect the function of the transporter associated with antigen processing, TAP. *PLoS Pathog.* **2008**, *4*, e1000080. [CrossRef] [PubMed]
66. Oettler, D.; Kaaden, O.R.; Neubauer, A. The equine herpesvirus 1 UL45 homolog encodes a glycosylated type II transmembrane protein and is involved in virus egress. *Virology* **2001**, *279*, 302–312. [CrossRef] [PubMed]
67. Kasem, S.; Yu, M.H.H.; Alkhalefa, N.; Ata, E.B.; Nayel, M.; Abdo, W.; Abdel-Moneim, A.S.; Fukushi, H. Impact of equine herpesvirus-1 ORF15 (EUL45) on viral replication and neurovirulence. *Vet. Microbiol.* **2024**, *298*, 110234. [CrossRef]
68. Wimer, C.L.; Schnabel, C.L.; Perkins, G.; Babasyan, S.; Freer, H.; Stout, A.E.; Rollins, A.; Osterrieder, N.; Goodman, L.B.; Glaser, A.; et al. The deletion of the ORF1 and ORF71 genes reduces virulence of the neuropathogenic EHV-1 strain Ab4 without compromising host immunity in horses. *PLoS ONE* **2018**, *13*, 11. [CrossRef]
69. Schnabel, C.L.; Wimer, C.L.; Perkins, G.; Babasyan, S.; Freer, H.; Watts, C.; Rollins, A.; Osterrieder, N.; Wagner, B. Deletion of the ORF2 gene of the neuropathogenic equine herpesvirus type 1 strain Ab4 reduces virulence while maintaining strong immunogenicity. *BMC Vet. Res.* **2018**, *14*, 245. [CrossRef]
70. Ata, E.B.; Zaghawa, A.; Ghazy, A.A.; Elsify, A.; Abdelrahman, K.; Kasem, S.; Nayel, M. Development and characterization of ORF68 negative equine herpes virus type-1, Ab4p strain. *J. Virol. Methods* **2018**, *261*, 121–131. [CrossRef]
71. Meindl, A.; Osterrieder, N. The equine herpesvirus 1 Us2 homolog encodes a nonessential membrane-associated virion component. *J. Virol.* **1999**, *73*, 3430–3437. [CrossRef]
72. Sharma, H.; Gulati, B.R.; Kapoor, S. Development of equine herpesvirus 1 persistently infected human lymphoblastoid cells expressing latency-associated transcripts. *Indian. J. Comp. Microb. Immunol. Infect. Dis.* **2020**, *41*, 117–124. [CrossRef]
73. Gao, J.; Liu, D.; Ma, Y.; Wang, X. The biology characteristics of herpesvirus latency associated transcripts. *Zhongguo Yufang Shouyi Xuebao/Chin. J. Prev. Vet. Med.* **2017**, *39*, 855–860.

74. Field, H.J.; Awan, A.R.; de la Fuente, R. Reinfection and reactivation of equine herpesvirus-1 in the mouse. *Arch. Virol.* **1992**, *123*, 409–419. [CrossRef] [PubMed]
75. Ahn, B.C.; Zhang, Y.; O'Callaghan, D.J. The equine herpesvirus-1 (EHV-1) IR3 transcript downregulates expression of the IE gene and the absence of IR3 gene expression alters EHV-1 biological properties and virulence. *Virology* **2010**, *402*, 327–337. [CrossRef] [PubMed]
76. Ahn, B.C.; Breitenbach, J.E.; Kim, S.K.; O'Callaghan, D.J. The equine herpesvirus-1 IR3 gene that lies antisense to the sole immediate-early (IE) gene is trans-activated by the IE protein, and is poorly expressed to a protein. *Virology* **2007**, *363*, 15–25. [CrossRef] [PubMed]
77. Holden, V.R.; Harty, R.N.; Yalamanchili, R.R.; O'Callaghan, D.J. The IR3 gene of equine herpesvirus type 1: A unique gene regulated by sequences within the intron of the immediate-early gene. *DNA Seq.* **1992**, *3*, 143–152. [CrossRef]
78. Dolan, A.; Jamieson, F.E.; Cunningham, C.; Barnett, B.C.; McGeoch, D.J. The genome sequence of herpes simplex virus type 2. *J. Virol.* **1998**, *72*, 2010–2021. [CrossRef]
79. Fang, J.C.; Liu, M.J. Translation initiation at AUG and non-AUG triplets in plants. *Plant Sci.* **2023**, *335*, 111822. [CrossRef]
80. de Arce, A.J.D.; Noderer, W.L.; Wang, C.L. Complete motif analysis of sequence requirements for translation initiation at non-AUG start codons. *Nucleic Acids Res.* **2018**, *46*, 985–994.
81. Harty, R.N.; Colle, C.F.; Grundy, F.J.; O'Callaghan, D.J. Mapping the termini and intron of the spliced immediate-early transcript of equine herpesvirus 1. *J. Virol.* **1989**, *63*, 5101–5110. [CrossRef]
82. Davison, A.J. Herpesvirus systematics. *Vet. Microbiol.* **2010**, *143*, 52–69. [CrossRef]
83. McGeoch, D.J.; Rixon, F.J.; Davison, A.J. Topics in herpesvirus genomics and evolution. *Virus Res.* **2006**, *117*, 90–104. [CrossRef]
84. Higgs, M.R.; Preston, V.G.; Stow, N.D. The UL15 protein of herpes simplex virus type 1 is necessary for the localization of the UL28 and UL33 proteins to viral DNA replication centres. *J. Gen. Virol.* **2008**, *89 Pt 7*, 1709–1715. [CrossRef] [PubMed]
85. Jacobson, J.G.; Yang, K.; Baines, J.D.; Homa, F.L. Linker insertion mutations in the herpes simplex virus type 1 UL28 gene: Effects on UL28 interaction with UL15 and UL33 and identification of a second-site mutation in the UL15 gene that suppresses a lethal UL28 mutation. *J. Virol.* **2006**, *80*, 12312–12323. [CrossRef] [PubMed]
86. Abbotts, A.P.; Preston, V.G.; Hughes, M.; Patel, A.H.; Stow, N.D. Interaction of the herpes simplex virus type 1 packaging protein UL15 with full-length and deleted forms of the UL28 protein. *J. Gen. Virol.* **2000**, *81 Pt 12*, 2999–3009. [CrossRef]
87. Mahmoud, H.Y.; Andoh, K.; Hattori, S.; Terada, Y.; Noguchi, K.; Shimoda, H.; Maeda, K. Characterization of glycoproteins in equine herpesvirus-1. *J. Vet. Med. Sci.* **2013**, *75*, 1317–1321. [CrossRef] [PubMed]
88. Oladunni, F.S.; Horohov, D.W.; Chambers, T.M. EHV-1: A Constant Threat to the Horse Industry. *Front. Microbiol.* **2019**, *10*, 2668. [CrossRef] [PubMed]
89. Smith, P.M.; Kahan, S.M.; Rorex, C.B.; von Einem, J.; Osterrieder, N.; O'Callaghan, D.J. Expression of the full-length form of gp2 of equine herpesvirus 1 (EHV-1) completely restores respiratory virulence to the attenuated EHV-1 strain KyA in CBA mice. *J. Virol.* **2005**, *79*, 5105–5115. [CrossRef]
90. von Einem, J.; Wellington, J.; Whalley, J.M.; Osterrieder, K.; O'Callaghan, D.J.; Osterrieder, N. The truncated form of glycoprotein gp2 of equine herpesvirus 1 (EHV-1) vaccine strain KyA is not functionally equivalent to full-length gp2 encoded by EHV-1 wild-type strain RacL11. *J. Virol.* **2004**, *78*, 3003–3013. [CrossRef]
91. Orbaum-Harel, O.; Sarid, R. Comparative Review of the Conserved UL24 Protein Family in Herpesviruses. *Int. J. Mol. Sci.* **2024**, *25*, 11268. [CrossRef]
92. Brito, A.F.; Baele, G.; Nahata, K.D.; Grubaugh, N.D.; Pinney, J.W. Intrahost speciations and host switches played an important role in the evolution of herpesviruses. *Virus Evol.* **2021**, *7*, 1. [CrossRef]
93. Davison, A.J. Evolution of the herpesviruses. *Vet. Microbiol.* **2002**, *86*, 69–88. [CrossRef]
94. Knizewski, L.; Kinch, L.; Grishin, N.V.; Rychlewski, L.; Ginalski, K. Human herpesvirus 1 UL24 gene encodes a potential PD-(D/E)XK endonuclease. *J. Virol.* **2006**, *80*, 2575–2577. [CrossRef]
95. Ruan, P.L.; Wang, M.S.; Cheng, A.C.; Zhao, X.X.; Yang, Q.; Wu, Y.; Zhang, S.Q.; Tian, B.; Huang, J.; Ou, X.M.; et al. Mechanism of herpesvirus UL24 protein regulating viral immune escape and virulence. *Front. Microbiol.* **2023**, *14*, 1268429. [CrossRef] [PubMed]
96. Bertrand, L.; Pearson, A. The conserved N-terminal domain of herpes simplex virus 1 UL24 protein is sufficient to induce the spatial redistribution of nucleolin. *J. Gen. Virol.* **2008**, *89 Pt 5*, 1142–1151. [CrossRef] [PubMed]
97. Bertrand, L.; Leiva-Torres, G.A.; Hyjazie, H.; Pearson, A. Conserved residues in the UL24 protein of herpes simplex virus 1 are important for dispersal of the nucleolar protein nucleolin. *J. Virol.* **2010**, *84*, 109–118. [CrossRef] [PubMed]
98. Lymberopoulos, M.H.; Pearson, A. Involvement of UL24 in herpes-simplex-virus-1-induced dispersal of nucleolin. *Virology* **2007**, *363*, 397–409. [CrossRef]
99. Carvalho, R.F.; Spilki, F.R.; Cunha, E.M.; Stocco, R.C.; Arns, C.W. Molecular data of UL24 homolog gene (ORF37) from Brazilian isolates of equine herpesvirus type 1. *Res. Vet. Sci.* **2012**, *93*, 494–497. [CrossRef]

100. Kasem, S.; Yu, M.H.; Yamada, S.; Kodaira, A.; Matsumura, T.; Tsujimura, K.; Madbouly, H.; Yamaguchi, T.; Ohya, K.; Fukushi, H. The ORF37 (UL24) is a neuropathogenicity determinant of equine herpesvirus 1 (EHV-1) in the mouse encephalitis model. *Virology* **2010**, *400*, 259–270. [CrossRef]
101. Hu, Y.; Zhang, S.Y.; Sun, W.C.; Feng, Y.R.; Gong, H.R.; Ran, D.L.; Zhang, B.Z.; Liu, J.H. Breaking Latent Infection: How ORF37/38-Deletion Mutants Offer New Hope against EHV-1 Neuropathogenicity. *Viruses* **2024**, *16*, 1472. [CrossRef]
102. Oguma, K.; Ishida, M.; Maeda, K.; Sentsui, H. Efficient Propagation of Equine Viruses in a Newly Established Equine Cell Line, FHK-Tcl3.1 Cells. *J. Vet. Med. Sci.* **2013**, *75*, 1223–1225. [CrossRef]
103. Andoh, K.; Kai, K.; Matsumura, T.; Maeda, K. Further Development of an Equine Cell Line that can be Propagated over 100 Times. *J. Equine Sci.* **2009**, *20*, 11–14. [CrossRef]
104. Maeda, K.; Yasumoto, S.; Tsuruda, A.; Andoh, K.; Kai, K.; Otoi, T.; Matsumura, T. Establishment of a novel equine cell line for isolation and propagation of equine herpesviruses. *J. Vet. Med. Sci.* **2007**, *69*, 989–991. [CrossRef]

veterinary sciences

MDPI

Article

Equine Distal Limb Wounds: Economic Impact and Short-Term Prognosis of Non-Synovial Versus Synovial Lesions in Southern Germany

Valeria Albanese [1], Paola Straticò [2,*], Holger Fischer [1] and Lucio Petrizzi [2]

1 Tierärztliches Kompetenzzentrum für Pferde Großwallstadt Altano GmbH, Niedernberger Str. 9, 63868 Großwallstadt, Germany; vzalbanese@gmail.com (V.A.); h.fischer@pferdeklinik-grosswallstadt.de (H.F.)
2 Department of Veterinary Medicine, Località Piano D'Accio, 64100 Teramo, Italy; lpetrizzi@unite.it
* Correspondence: pstratico@unite.it

Simple Summary: Injuries to the limbs are a common occurrence in horses. Sometimes joints, bursas, or tendon sheaths, so called synovial structures, may be involved in limb wounds, too. This carries the risk of infection of such structures, which may have a clinical and financial impact. Additionally, tendons and/or ligaments may be involved. The aim of this study is to compare wounds with and without involvement of synovial structures, and wounds with and without tendon and/or ligament involvement, as far as cost of treatment and clinical outcomes.

Abstract: Injuries to the distal limb are common in horses. The clinical aspect of the lesions is variable based on the structures that are involved. Synovial structures as well as tendons and/or ligaments may be involved in such injuries, affecting treatment modalities and costs, and prognosis. This retrospective study compares wounds involving synovial structures to wounds without such involvement in terms of treatment duration and costs, and prognosis. Synovial wounds were much more expensive to treat and carried a lower prognosis. Involvement of tendons and/or ligaments in the wounds resulted in more expensive and prolonged treatment courses compared to cases without such involvement, even without synovial contamination. The prognosis for discharge was good overall for both groups.

Keywords: wound; distal limb; synovial cavity; synovial infection; horses

1. Introduction

Injuries to the distal limb are common in horses. The clinical aspect of the lesions is variable based on which structures are involved. The anatomy of the musculotendinous apparatus of equine distal limb is made complex by the presence of synovial and non-synovial structures very close to each other and poorly protected by surrounding tissues [1]. The involvement of either of the two, or both, affects the prognosis as well as the type of intervention. The presence and proliferation of bacteria within a synovial cavity leads to severe inflammation and secretion of inflammatory mediators that directly cause modification of the internal homeostasis of the injured synovial cavity, and metabolic changes with secondary damage to the articular surface or teno-ligamentous structures, with secondary osteoarthritis and adhesions [1,2]. Early recognition of involved structures is mandatory to establish an appropriate treatment and improve prognosis [3–5].

A multimodal approach is suggested when a synovial lesion is suspected. Treatment aims to remove pathogen and inflammatory mediators [6]. Also, non-synovial wounds

 https://doi.org/10.3390/vetsci12030205

need to have a prompt intervention, to avoid migration of the microorganism into the adjacent synovial cavity [7]. Systemic broad-spectrum antibiotics are usually combined with local intravenous limb perfusion (IVLP). Surgical debridement and synovial lavage are recommended to remove debris and devitalized tissue, and enhance vascularization and local antibiotic diffusion [8,9].

Prognosis is usually good for survival: according to published literature, 85% of horses with synovial infections survive to discharge. The percentage of success decreases when return to athletic function is evaluated [10–13].

How quickly treatment is initiated is deemed important, with best prognoses associated with a time frame of 24–36 h from the initial trauma [11,14,15]. Economic constraints may play a role in the decision process on what treatment is affordable, with direct effects on the prognosis.

This retrospective study was conducted to investigate clinical outcomes of horses referred for limb wounds for which surgical treatment was elected.

The primary objective was to assess and compare the clinical and economic impact of synovial versus non-synovial limb wounds in horses. Key metrics included duration of hospitalization, total and daily hospitalization costs, number of surgeries, duration of antibiotic treatment, and survival to discharge. Secondary objectives were to differentiate outcomes between complicated and non-complicated cases within the non-synovial group, to compare the economic impact of complicated non-synovial wounds to those involving one or more synovial structures, and to examine the association of the use of IVLP with survival in synovial cases. (Table 1).

Table 1. Objectives of the study.

1. Objective	2. Objective	3. Objective	4. Objective
Synovial vs. non-synovial limb wounds	Complicated vs. non-complicated non-synovial wounds	Complicated non-synovial wounds vs. synovial wounds	Association of IVLP with survival in synovial cases

2. Materials and Methods

2.1. Case Definition and Selection

Study population consisted of horses referred to a private equine hospital located in southern Germany for limb wounds between 2021 and 2023. Inclusion criteria were: age older than 1 year, one or multiple limb lacerations, surgical treatment under general anesthesia. Additional injuries such as fractures, or tendon or ligament transections of subjectively more than 25% of the anatomical structure's cross-section, were considered exclusion criteria.

Horses were classified into two cohorts, as determined by clinical evaluations and diagnostics. Cohort 1 (synovial) included horses with wounds involving synovial structures. All horses in Cohort 1 underwent one or more arthroscopic lavage. Cohort 2 (non-synovial) included horses with wounds not involving synovial structures. Cohort 2 was further divided into complicated and non-complicated cases. Cases were defined as complicated when one or multiple tendons and/or ligaments were involved in the laceration.

2.2. Data Collection

Data were retrospectively collected from medical records. Signalment, time to referral, and location of the wound were recorded. The evaluated variables included duration of hospitalization, total and daily cost of hospitalization, number of surgical procedures performed, duration of antibiotic treatment, use of intravenous limb perfusion (IVLP), and survival to discharge. The use of intravenous regional limb perfusion in Cohort 1 and its association with survival outcomes were also analyzed. Additionally, time elapsed between wound detection and referral (time to referral) was examined across both cohorts to assess its association with survival, cost, duration of hospitalization, and number of surgical procedures.

2.3. Statistical Analysis

Comparisons were made between synovial and non-synovial groups and between non-synovial groups for complicated versus non-complicated cases. Fisher's exact test and a Pearson's Chi-squared test were used to investigate categoric variables. A Wilcoxon rank sum test was used to compare groups for continuous variables. The association between survival to discharge and intravenous regional limb perfusion was evaluated using the Chi-square test. A linear and logistic regression assessed the impact of the cost per day, hospitalization length, and the number of surgeries on time to referral for both cohorts, respectively, for continuous and categorical variables. Statistical significance was set for $p < 0.05$.

3. Results

A total of 51 horses were included in the study. For 21/51 (41%) of them, involvement of at least one synovial structure was confirmed. These horses were allocated to Cohort 1. Thirty (59%) did not show any synovial involvement; therefore, these horses were allocated to Cohort 2.

In the synovial cohort, 24 structures were involved in 21 horses (Table 2). In three horses, multiple structures were involved.

Table 2. Synovial structures involved in Cohort 1.

Structure	Number
Digital flexor tendon sheath, hind	7
Tibio-tarsal joint	4
Metatarsophalangeal joint	2
Metacarpophalangeal joint	2
Radial carpal joint	2
Digital flexor tendon sheath, fore	1
Navicular bursa, fore	1
Elbow	1
Calcaneal bursa	1
Distal intertarsal joint	1
Tarsometatarsal joint	1
Tarsal sheath	1
Total	**24**

All 30 horses in the non-synovial cohort survived to discharge (100%). Of the 21 horses in the synovial cohort, 18 survived to discharge (85.7%).

In Table 3, the differences between the two Cohorts in the number of surgeries, duration of antibiotic treatment, total hospitalization costs, length of hospitalization, cost per day, and survival to discharge are summarized.

Table 3. Comparative analysis between synovial vs. non-synovial joint involvement cases.

	Non-Synovial, n = 30 [1]	Synovial, n = 21 [1]	p-Value [2]
Number of Surgeries			
1	28 (93.3%)	15 (71.4%)	0.040
2	2 (6.7%)	3 (14.3%)	
3	0 (0%)	3 (14.3%)	
Duration of Antibiotic Treatment (days)	16 (12–19)	20 (17–25)	0.018
Total Hospitalization Cost (euro)	4.384 (2.746–6.489)	5.979 (5.050–9.662)	0.003
Length of Hospitalization (days)	10 (7, 13)	11 (8, 17)	0.3
Cost Per Day (euro)	403 (330–499)	536 (424–652)	0.030
Survival to Discharge (number of horses)	30 (100%)	18 (85.7%)	0.064

[1] n (%); Median (IQR), [2] Fisher's exact test; Pearson's Chi-squared test; Wilcoxon rank sum test; Wilcoxon rank sum exact test.

There were notable differences in hospitalization metrics within Cohort 2, when comparing complicated and non-complicated cases (Table 4): a significantly longer median length of hospitalization for complicated cases, and a higher hospitalization cost.

Table 4. Non-synovial group complicated versus not complicated length of hospitalization, hospitalization total and daily cost analysis.

	Complicated, n = 5 [1]	Not Complicated, n = 25 [1]	p-Value [2]
Length of Hospitalization (days)	24 (13, 26)	10 (7, 12)	**0.024**
Total Hospitalization Cost (Euro)	8000 (6494–8199)	3834 (2675–5041)	**0.016**
Cost Per Day (Euro)	342 (308–475)	405 (346–500)	0.6

[1] Median (IQR); [2] Wilcoxon rank sum test; Wilcoxon rank sum exact test.

Time to referral ranged between 6 and 42 days, with a median value of 7 days (with a median time for both cohorts). Univariate regression analysis was employed to investigate how different outcomes were associated with time elapsed between diagnosis and referral (time to referral) in both cohorts (Table 5). In Cohort 1, there was a significant association found between time to referral and cost per day (coefficient: 0.94, $p < 0.001$). This is a relationship that was not observed in Cohort 2. No significant association between time to referral and hospitalization length, or between time to referral and number of surgeries was noted in either group (Table 6).

Table 5. Univariate regression analysis for association of different factors with the time to referral for synovial and non-synovial cases.

Variable	Cohort 1			Cohort 2		
	Coefficient [1]	95% CI	p-Value	Beta [1]	95% CI	p-Value
Cost per Day	0.94	0.52, 1.4	<0.001	1.0	−1.5, 3.5	0.4
Total Hospitalization Cost	−0.49	−7.4, 6.5	0.9	−0.52	−27, 26	>0.9
Hospitalization Length	−0.03	−0.09, 0.04	0.5	−0.03	−0.13, 0.07	0.5
Number of Surgeries	0.00	−0.01, 0.00	0.5	0.01	−0.09, 0.05	0.8

CI = Confidence Interval. Note: Linear regression was used for the cost per day, total hospitalization cost, and hospitalization length (coefficient = Beta) while logistic regression was used for number of surgeries (coefficient = Odds ratio). [1] Coefficient is the Odd Ratio resulting from the the linear or logistic regression. Beta is the same.

Table 6. Chi-Square Test for association of survival-to-discharge with intravenous regional limb perfusion [1].

	Chi-Square Test for Association Survival to Discharge and IVRLP	
	IVRLP Given [2]	IVRLP Not Given [2]
Survivors (number of horses)	0	3
Non-survivors (number of horses)	11	7

[1] Chi-square statistic = 1.79, df = 1, p-value = 0.181. [2] Pearson's Chi-squared test with Yates' continuity correction.

The study also examined the association of survival to discharge with the use of IVLP in synovial cases. A Chi-square test found no significant association between these variables ($p = 0.181$) (Table 6).

A reliable analysis between 'survival to discharge' and 'duration prior to referral' was not possible due to a significant imbalance of data. For the same reason, it was not possible to evaluate possible differences in prognoses within Cohort 1 related to which structure was involved.

4. Discussion

4.1. Multiple Surgical Procedures

Contamination of one or multiple synovial structures may warrant multiple surgical procedures to optimize the chances of a full recovery. According to the previous literature [16], contaminated or infected synovial structures need to be lavaged promptly, aggressively, thoroughly and, at times, repeatedly. In this dataset, a higher percentage of horses with synovial involvement (28.6%) needed multiple procedures than horses without such involvement (6.7%). Multiple procedures consisting in joint lavages were deemed necessary to address synovial contamination in those cases. Although a multimodal approach to the treatment of wounds other than antimicrobials in horses is described [17], in this cohort study the antibiotic therapy still showed a high impact on the medical strategy, with longer treatment in the synovial cohort.

A thorough report and analysis of systemic antimicrobial treatment protocols go beyond the purpose of this work. However, it is worth mentioning that most horses in this report received an initial course of 5–7 days broad-spectrum antibiotic combination (Procain penicillin 22,000 IU/kg IM BID or Amoxicillin 10 mg/kg IV BID, and Gentamicin 6.6 mg/kg IV SID) followed by a course of oral antibiotic (Trimethoprim-Sulfadiazin

30 mg/kg PO BID or Doxycyclin 10 mg/kg PO BID or Enrofloxacin 7.5 mg/kg PO SID), depending on the results of culture sensitivity tests. Treatment was administered if it was deemed clinically necessary for each individual case.

Most horses with synovial involvement (71.4%) only needed one lavage to resolve. Synovial lavage was performed via arthroscopy, tenoscopy, or bursoscopy. Compared to through-and-through needle lavage, the endoscopic technique offers many advantages. The operator may benefit from visualization of the affected structure and foreign material, and fibrin pannus or debris may be removed with instruments. In addition, higher fluid volume and pressure can be utilized. Disadvantages of arthroscopic technique are the need of specialized instrumentation and skills, as well as the costs of the procedure [2,12,18,19]. The choice of arthroscopic lavage over needle through and through lavage may have also contributed to treatment success after only one procedure.

Furthermore, referral and subsequent treatment were mostly carried out promptly after diagnosis. Therefore, most affected synovial structures were likely lavaged before contamination turned into an established infection, which may have also contributed to the success rate of only one lavage procedure.

4.2. Time to Referral

The literature suggests that timely intervention leads to a better prognosis in cases of synovial contamination; the past literature has mostly supported this claim [8,15,20,21]. Contamination of a synovial structure does not equal infection of said synovial structure based on the presence or proliferation of bacteria [22]. However, untreated contamination may turn into an infection and affect the prognosis [10–12,23,24].

No additional diagnostics were consistently carried out to differentiate between contaminated and infected structures in our data.

The distribution of data made a statistical evaluation of the association between time to referral and survival not possible. Nevertheless, in both cohorts the survival rate was close to 100% (85.7% in Cohort 1; 100/in Cohort 2) despite the wide range of time to refer.

A significant positive association was observed between the time to referral and daily costs of treatment ($p < 0.001$) for horses in the synovial cohort. This association was not significant for non-synovial cases, where the impact of time to referral on daily costs was not significant ($p = 0.4$).

There may be several reasons for this. The initial assessment of a wound with suspected synovial involvement may be more expensive [25,26] in terms of time and diagnostics. The use of larger volumes of fluids while lavaging a synovial structure with an infection of several days' duration may also result in higher costs. Unfortunately, a breakdown of the bill was not performed to further investigate these possibilities. In any case, this finding indicates that a longer waiting time before referral could be economically detrimental. Many times, the very reason for delaying referral is, indeed, the perception that there would be a more thrifty and just as successful treatment modality in the field. However, if referral is a necessity, delaying it may end up in even higher veterinary costs.

The lack of correlation between time to referral and length of hospitalization ($p = 0.5$), as well as the lack of correlation between time to referral and the number of surgical procedures performed ($p = 0.5$ for synovial cohort, $p = 0.8$ for non-synovial cohort) align with the previous literature claiming that delaying time to referral may not have a clinically negative impact on either of them [13,19,22,27,28]. The lack of impact of time to referral on these parameters reflects that the chosen treatment defeated the infection just as effectively, irrespectively of the duration of the infection itself. This could be due to the type of bacteria involved and their antibiotic susceptibility. All lacerations occurred at home, and not in a hospital setting. Therefore, infections were most likely caused by wild-type bacteria, for which antibiotic resistance is not as common as it is for nosocomial bacteria [29]. Another possible reason is the aggressive initial surgical debridement; all horses involved in this study underwent major surgical wound debridement with or without synovial lavage. The debridement could have been radical enough to excise most infected tissue even in long standing cases.

4.3. *Length of Hospitalization*

This population of horses consists predominantly of low- and medium-level sport horses, owned by non-professional equestrians. Many horses in the area are kept in paddocks or pastures and cannot be confined to a stall or box at home. Although no data are available, it is the authors' opinion that the combination of the lack of sufficient clients' knowledge and expertise and the lack of facilities to confine convalescent horses in at home may lead to longer hospitalizations.

Horses in both cohorts were hospitalized for a similar length of time (range 7–13 days, median 10 days for the non-synovial cohort; range 8–17 days, median 11 days for the synovial cohort). One of the factors that leads to this finding may be that most horses in the synovial cohort only required one surgery. On the other hand, the lack of significant difference could also be because horses in the non-synovial cohort were hospitalized longer than strictly necessary to accommodate owner's needs and preferences.

4.4. *Treatment Cost*

The total cost of treatment was significantly higher for horses in the synovial cohort (range EUR 5050–9662, median EUR 5979) than for horses in the non-synovial cohort (range EUR 2746–6489, median EUR 4384) ($p = 0.003$). Daily costs of treatment were significantly higher for the synovial than for the non-synovial cohort. Factors involved in this difference were higher surgical fees for the arthroscopic lavage compared to wound debridement and care, as well as material used and level of post-operative care and diagnostics, such as repeated synoviocentesis to help assess recovery.

4.5. *Prognosis*

All horses in the non-synovial cohort and most horses in the synovial cohort survived to discharge. Three horses in the synovial group were euthanized before discharge. One had a wound that caused an infection of the elbow joint; the patient was referred 42 days after the wound occurred, and, despite aggressive repeated lavage, remained lame, and was, therefore, euthanized. The second non-surviving horse had both distal intertarsal and tarsometatarsal joints involved. Despite IVLP and needle through-and-through lavage of the joints performed twice under general anesthesia, the horse's level of comfort deteriorated, and was therefore euthanized. Lower hock joints are not amenable to arthroscopic lavage due to their anatomy. Therefore, a through-and-through needle lavage was performed. This technique allows only limited amounts of fluids through the joint space and, since an instrument portal is not present, does not allow for debridement of any necrotic material or pannus. Therefore, it is less effective in removing all the infection from the joint space, which might have been the reason for failure in this case. The third non-survivor was a yearling that, after successful treatment of a tarso-crural sepsis following a laceration, developed a septic physitis in another limb and was euthanized. The pathogenesis of septic physitis was most likely hematogenous spread from the septic joint. Treatment could have been implemented, consisting in intravenous regional limb perfusion and direct antibiotic treatment of the physis self; however, any further treatment was declined.

All three non-survivors had concurrent factors that made their clinical course inherently more complicated than a simple laceration with synovial involvement: in one case the wound was over 6 weeks old. In the second case, the joint was poorly accessible for treatment. In the third case, an additional, most likely hematogenous infection presented itself at a different, remote site.

Treatment of septic synovitis in horses has very broad survival rates ranging from between 56% and 100% [9,10,13,15,19–21,28,30–36]. The prognosis for short-term survival in the synovial cohort (89%) is at the high end of that range. Nevertheless, the prognosis for survival of the synovial cohort was lower than that of the non-synovial cohort (100%). The difference approximated but did not quite reach statistical significance (p = 0.064). These data are in accordance with previous studies [37], despite the potential selection bias that this study could have suffered. In fact, only horses undergoing surgery were enrolled, leading to the exclusion of those horses that have been euthanized before referral for economic of perceived poor prognosis. This may have overestimated the survival rate and therefore the prognosis.

4.6. Comparison Complicated Non Synovial/Synovial

The median cost of treatment for cases without synovial involvement but with tendons and/or ligaments lesions was significantly higher than for cases without such involvement, indicating a substantial financial impact of such finding.

The number of surgical procedures under general anesthesia and the total length of hospitalization were not significantly different among the two subsets.

This finding is especially relevant at the time of formulating an estimate in terms of time and costs before undertaking treatment. It appears, then, that the involvement of tendons and ligaments contributes to much higher costs and longer treatment.

Unfortunately, due to the retrospective nature of the study, data about the return to athletic function was missing in too many cases to render statistical analysis of this variable not possible.

Additional limitations of this study are its small sample size, grouping of all synovial structures in one cohort, lack of some data such as culture and sensitivity results and return to function, and its possible aforementioned selection bias.

5. Conclusions

Traumatic wounds of the equine distal limb in horses carry a good prognosis for survival in the geographic area where the study was performed. This may be partly due to management practices and referral practices in southern Germany and may not necessarily reflect the reality of other geographical areas. In the case of synovial involvement and non-synovial lesions involving a tendon or a ligament, owners in this area must be informed about the high treatment costs. In the case of synovial lesions, a multimodal treatment based on synovial lavage under general anesthesia with arthroscopic guidance, and systemic and local antibiotics must be set up as soon as possible. When the lesion is not penetrating a synovial structure, surgical debridement and antimicrobial therapy should be advised. Although no association between time to referral and prognosis was found in this research, prompt intervention and referral are advisable to reduce the chances of an infection developing.

Author Contributions: Conceptualization, V.A. and P.S.; methodology, V.A.; investigation, V.A.; resources, H.F.; data curation, V.A.; writing—original draft preparation, V.A.; writing—review and editing, P.S. and L.P.; funding acquisition, H.F. All authors have read and agreed to the published version of the manuscript.

Funding: This research was privately funded.

Institutional Review Board Statement: Ethical review and approval were waived for this study due to the retrospective and observational nature of this study. Informed consent was obtained from all clients for the use of their animals' medical records for research purposes.

Informed Consent Statement: Informed consent was obtained from all subjects involved in the study.

Data Availability Statement: Data supporting the findings of this study are stored securely at the institution. Due to confidentiality agreements and ethical restrictions, the raw data are not publicly available but can be provided upon reasonable request to the corresponding author.

Conflicts of Interest: Author Valeria Albanese and Holger Fischer were employed by Tierärztliches Kompetenzzentrum für Pferde Großwallstadt Altano GmbH. The funders had no role in the design of the study; in the collection, analyses, or interpretation of data; in the writing of the manuscript; or in the decision to publish the results.

References

1. Baxter, G.M. Management of Wounds Involving Synovial Structures in Horses. *Clin. Tech. Equine Pract.* **2004**, *3*, 204–214. [CrossRef]
2. Joyce, J. Injury to Synovial Structures. *Vet. Clin. N. Am. Equine Pract.* **2007**, *23*, 103–116. [CrossRef] [PubMed]
3. Ludwig, E.K.; van Harreveld, P.D. Equine Wounds over Synovial Structures. *Vet. Clin. Equine Pract.* **2018**, *34*, 575–590. [CrossRef] [PubMed]
4. Michotte, M.; Raes, E.; Oosterlinck, M. Diagnostic Accuracy of Plain Radiography to Identify Synovial Penetration in Horses with Traumatic Limb Wounds. *Equine Vet. J.* **2024**, *57*, 62–68. [CrossRef] [PubMed]
5. Bryant, H.A.; Dixon, J.J.; Weller, R.; Bolt, D.M. Use of Positive Contrast Radiography to Identify Synovial Involvement in Horses with Traumatic Limb Wounds. *Equine Vet. J.* **2019**, *51*, 20–23. [CrossRef]
6. Ludwig, E.K.; Wiese, R.B.; Graham, M.R.; Tyler, A.J.; Settlage, J.M.; Were, S.R.; Petersson-Wolfe, C.S.; Kanewsky-Mullarky, I.; Dahlgren, L.A. Serum and Synovial Fluid Serum Amiloid A Response in Equine Models of Synovitis and Septic Arthritis. *Vet. Surg.* **2016**, *45*, 859–867. [CrossRef] [PubMed]
7. Carstanjen, B.; Boehart, S.; Cislakova, M. Septic Arthritis in Adult Horses. *Pol. J. Vet. Sci.* **2010**, *13*, 201. [PubMed]
8. Schneider, R.K.; Bramlage, L.R.; Mecklenburg, L.M.; Moore, R.M.; Gabel, A.A. Open Drainage, Intra-articular and Systemic Antibiotics in the Treatment of Septic Arthritis/Tenosynovitis in Horses. *Equine Vet. J.* **1992**, *24*, 443–449. [CrossRef]
9. Cousty, M.; David Stack, J.; Tricaud, C.; David, F. Effect of Arthroscopic Lavage and Repeated Intra-articular Administrations of Antibiotic in Adult Horses and Foals with Septic Arthritis. *Vet. Surg.* **2017**, *46*, 1008–1016. [CrossRef] [PubMed]
10. Schneider, R.K.; Bramlage, L.R.; Moore, R.M.; Mecklenburg, L.M.; Kohn, C.W.; Gabel, A.A. A Retrospective Study of 192 Horses Affected with Septic Arthritis/Tenosynovitis. *Equine Vet. J.* **1992**, *24*, 436–442. [CrossRef] [PubMed]
11. Gibson, K.T.; McIlwraith, C.W.; Turner, A.S.; Stashak, T.S.; Aanes, W.A.; Trotter, G.W. Open Joint Injuries in Horses: 58 Cases (1980–1986). *J. Am. Vet. Med. Assoc.* **1989**, *194*, 398–404. [CrossRef]
12. Seabaugh, K.A.; Baxter, G.M. Diagnosis and Management of Wounds Involving Synovial Structures. In *Equine Wound Management*, 3rd ed.; Wiley: Hoboken, NJ, USA, 2008; pp. 463–488.
13. Walmsley, E.A.; Anderson, G.A.; Muurlink, M.A.; Whitton, R.C. Retrospective Investigation of Prognostic Indicators for Adult Horses with Infection of a Synovial Structure. *Aust. Vet. J.* **2011**, *89*, 226–231. [CrossRef]
14. Wright, I.M.; Scott, M. Management of Penetrating Wounds in Joints, Tendon Sheaths and Bursae. *Equine Vet. Educ.* **1989**, *1*, 15–22. [CrossRef]
15. Fraser, B.S.L.; Bladon, B.M. Tenoscopic Surgery for Treatment of Lacerations of the Digital Flexor Tendon Sheath. *Equine Vet. J.* **2004**, *36*, 528–531. [CrossRef]
16. Ahern, B.; Richardson, D. Surgical Site Infections and the Use of Antimicrobials. In *Equine surgery*, 4th ed.; Auer, J.A., Stick, J.A., Eds.; Elsevier Saunders: St. Louis, MO, USA, 2012; pp. 68–84. [CrossRef]
17. Freeman, S.L.; Ashton, N.M.; Elce, Y.A.; Hammond, A.; Hollis, A.R.; Quinn, G. BEVA Primary Care Clinical Guidelines: Wound Management in the Horse. *Equine Vet. J.* **2021**, *53*, 18–29. [CrossRef] [PubMed]
18. Morton, A.J. Diagnosis and Treatment of Septic Arthritis. *Vet. Clin. Equine Pract.* **2005**, *21*, 627–649. [CrossRef]
19. Wright, I.M.; Smith, M.R.W.; Humphrey, D.J.; Eaton-Evans, T.C.J.; Hillyer, M.H. Endoscopic Surgery in the Treatment of Contaminated and Infected Synovial Cavities. *Equine Vet. J.* **2003**, *35*, 613–619. [CrossRef] [PubMed]
20. Wereszka, M.M.; White, N.A.; Furr, M.O. Factors Associated with Outcome Following Treatment of Horses with Septic Tenosynovitis: 51 Cases (1986–2003). *J. Am. Vet. Med. Assoc.* **2007**, *230*, 1195–1200. [CrossRef]

21. Findley, J.A.; Pinchbeck, G.L.; Milner, P.I.; Bladon, B.M.; Boswell, J.; Mair, T.S.; Suthers, J.M.; Singer, E.R. Outcome of Horses with Synovial Structure Involvement Following Solar Foot Penetrations in Four UK Veterinary Hospitals: 95 Cases. *Equine Vet. J.* **2014**, *46*, 352–357. [CrossRef] [PubMed]
22. Rubio-Martínez, L.M.; Cruz, A.M. Antimicrobial Regional Limb Perfusion in Horses. *J. Am. Vet. Med. Assoc.* **2006**, *228*, 706–712. [CrossRef]
23. Honnas, C.M.; Schumacher, J.; Cohen, N.D.; Watkins, J.P.; Taylor, T.S. Septic Tenosynovitis in Horses: 25 Cases (1983–1989). *J. Am. Vet. Med. Assoc.* **1991**, *199*, 1616–1622. [CrossRef]
24. Furness, M.C.; Snyman, H.N.; Abrahams, M.; Moore, A.; Vince, A.; Anderson, M.E.C. Severe Gastric Impaction Secondary to a Gastric Polyp in a Horse. *Can. Vet. J.* **2013**, *54*, 979–982. [PubMed]
25. Ross, M.W.; Orsini, J.A.; Richardson, D.W.; Martin, B.B. Closed Suction Drainage in the Treatment of Infectious Arthritis of the Equine Tarsocrural Joint. *Vet. Surg.* **1991**, *20*, 21–29. [CrossRef]
26. Ooi, C.C.; Schneider, M.E.; Malliaras, P.; Chadwick, M.; Connell, D.A. Diagnostic Performance of Axial-Strain Sonoelastography in Confirming Clinically Diagnosed Achilles Tendinopathy: Comparison with B-Mode Ultrasound and Color Doppler Imaging. *Ultrasound Med. Biol.* **2015**, *41*, 15–25. [CrossRef]
27. Milner, P.I.; Bardell, D.A.; Warner, L.; Packer, M.J.; Senior, J.M.; Singer, E.R.; Archer, D.C. Factors Associated with Survival to Hospital Discharge Following Endoscopic Treatment for Synovial Sepsis in 214 Horses. *Equine Vet. J.* **2014**, *46*, 701–705. [CrossRef]
28. Smith, L.J.; Mellor, D.J.; Marr, C.M.; Mair, T.S. What Is the Likelihood That a Horse Treated for Septic Digital Tenosynovitis Will Return to Its Previous Level of Athletic Function? *Equine Vet. J.* **2006**, *38*, 337–341. [CrossRef] [PubMed]
29. Sousa, S.A.; Feliciano, J.R.; Pita, T.; Soeiro, C.F.; Mendes, B.L.; Alves, L.G.; Leitao, J.H. Bacterial Nosocomial Infections: Multidrug Resistance as a Trigger for the Development of Novel Antimicrobials. *Antibiotics* **2021**, *10*, 942. [CrossRef]
30. Stewart, A.A.; Goodrich, L.R.; Byron, C.R.; Evans, R.B.; Stewart, M.C. Antimicrobial Delivery by Intrasynovial Catheterisation with Systemic Administration for Equine Synovial Trauma and Sepsis. *Aust. Vet. J.* **2010**, *88*, 115–123. [CrossRef] [PubMed]
31. Smith, L.J.; Marr, C.M.; Payne, R.J.; Stoneham, S.J.; Reid, S.W.J. What Is the Likelihood That Thoroughbred Foals Treated for Septic Arthritis Will Race? *Equine Vet. J.* **2004**, *36*, 452–456. [CrossRef]
32. Post, E.M.; Singer, E.R.; Clegg, P.D.; Smith, R.K.; Cripps, P.J. Retrospective Study of 24 Cases of Septic Calcaneal Bursitis in the Horse. *Equine Vet. J.* **2003**, *35*, 662–668. [CrossRef]
33. Meijer, M.C.; Van Weeren, P.R.; Rijkenhuizen, A.B.M. Clinical Experiences of Treating Septic Arthritis in the Equine by Repeated Joint Lavage: A Series of 39 Cases. *J. Vet. Med. Ser. A* **2000**, *47*, 351–365. [CrossRef]
34. Vos, N.J.; Ducharme, N.G. Analysis of Factors Influencing Prognosis in Foals with Septic Arthritis. *Ir. Vet. J.* **2008**, *61*, 102. [CrossRef] [PubMed]
35. Johnson, S.A.; Biscoe, E.W.; Eilertson, K.E.; Lutter, J.D.; Schneider, R.K.; Roberts, G.D.; Cary, J.A.; Frisbie, D.D. Tissue Predictability of Elastography Is Low in Collagenase Induced Deep Digital Flexor Tendinopathy. *Vet. Radiol. Ultrasound* **2022**, *63*, 111–123. [CrossRef]
36. Steel, C.M.; Hunt, A.R.; Adams, P.L.E.; Robertson, I.D.; Chicken, C.; Yovich, J.V.; Stick, J.A. Factors Associated with Prognosis for Survival and Athletic Use in Foals with Septic Arthritis: 93 Cases (1987–1994). *J. Am. Vet. Med. Assoc.* **1999**, *215*, 973–977. [CrossRef] [PubMed]
37. de Souza, T.C.; Burford, J.; Busschers, E.; Freeman, S.; Suthers, J.M. Multicenter Study Investigating Long-Term Survival After Synovial Lavage of Contaminated and Septic Synovial Structures in Horses Presented to 10 UK Referral Hospitals. *Vet. Surg.* **2024**, *53*, 1083–1092. [CrossRef] [PubMed]

veterinary sciences

Article

Aflatoxin B1-Induced Apoptosis in Donkey Kidney via EndoG-Mediated Endoplasmic Reticulum Stress

Yanfei Ji, Yu Zhang, Wenxuan Si, Jing Guo, Guiqin Liu, Changfa Wang, Muhammad Zahoor Khan, Xia Zhao * and Wenqiang Liu *

College of Agriculture and Biology, Liaocheng University, Liaocheng 252000, China
* Correspondence: zhaoxia@lcu.edu.cn (X.Z.); liuwenqiang@lcu.edu.cn (W.L.)

Simple Summary: This study investigates the effects of Aflatoxin B1 (AFB1) on kidney damage in donkeys, aiming to uncover the molecular mechanisms and pathways responsible for nephrotoxicity. Aflatoxin B1 is a common environmental contaminant known to pose significant health risks to both humans and livestock, with nephrotoxicity being one of its most prominent toxicological effects. In this study, a donkey model was exposed to AFB1, resulting in observable kidney damage, apoptosis, and oxidative stress. The findings highlight the upregulation of Endonuclease G (EndoG), which plays a crucial role in triggering endoplasmic reticulum (ER) stress and activating mitochondrial apoptosis. These results provide valuable insights into the molecular mechanisms underlying AFB1-induced nephrotoxicity in donkeys.

Abstract: Aflatoxin B1 (AFB1) is a prevalent environmental and forage contaminant that poses significant health risks to both humans and livestock due to its toxic effects on various organs and systems. Among its toxicological effects, nephrotoxicity is a hallmark of AFB1 exposure. However, the precise mechanisms underlying AFB1-induced kidney damage in donkeys remain poorly understood. To investigate this, we established a donkey model exposed to AFB1 by administering a diet supplemented with 1 mg AFB1/kg for 30 days. Kidney apoptosis was assessed using TUNEL staining, while gene expression and protein levels of Endonuclease G (EndoG), as well as genes related to endoplasmic reticulum (ER) stress and apoptosis, were quantified by RT-qPCR and Western blotting. Our findings indicate that AFB1 exposure resulted in significant kidney injury, apoptosis, and oxidative stress. Notably, AFB1 exposure upregulated the expression of EndoG and promoted its translocation to the ER, which subsequently induced ER stress and activated the mitochondrial apoptotic pathway. These results suggest that AFB1-induced kidney damage in donkeys is mediated through the oxidative stress and mitochondrial apoptosis pathways, primarily involving the EndoG-IRE1/ATF6-CHOP signaling axis.

Keywords: AFB1; donkey; kidney; nephrotoxicity; EndoG; endoplasmic reticulum stress; apoptosis

1. Introduction

Aflatoxin is a type of mycotoxin primarily produced by Aspergillus flavus and Aspergillus parasiticus, with different chemotypes that includes aflatoxin B1 (AFB1), B2, B2a, G1, G2, G2a, M1, and M2. Among these, AFB1 is considered the most toxic and highly carcinogenic [1–3]. AFB1 is commonly found in the environment, soil, food crops, and animal feed. The consumption of contaminated feed can lead to livestock disease and even death [4,5]. Over 80% of ingested AFB1 is absorbed in the duodenum and jejunum through passive transport. It then accumulates in the liver and kidneys. The metabolites of AFB1 are primarily excreted through the kidneys, resulting in a relatively high residual amount of AFB1 in the kidneys. This can cause renal dysfunction and damage [6]. Research has reported that oxidative stress is a common and significant mechanism in the toxicology of AFB1 [7]. AFB1 disrupts the antioxidant system in mammalian cells and induces the overproduction of reactive oxygen species (ROS), which triggers oxidative stress damage and other signaling cascades, ultimately resulting in cell death [8–10]. Studies have also shown that chronic AFB1 intoxication can cause various injuries in livestock and poultry, including anorexia, liver necrosis, gallbladder enlargement, and intestinal congestion. These injuries occur through multiple biological processes, such as inflammation, apoptosis, and programmed death [11–15]. Additionally, AFB1 can cause renal injury by activating oxidative stress-related signaling pathways and apoptosis [16].

Endonuclease G (EndoG) is a nuclear-coded protein located in the intermembrane space of mitochondria. When subjected to external stimulation, EndoG can translocate to the endoplasmic reticulum (ER) and bind to Bip (also named GRP78), resulting in the release of IRE1a/PERK and activation of the ER stress response [17]. The ER plays a crucial role in cellular protein fold and quality control and Ca^{2+} storage and release, which are essential for cell survival [18]. Protein folding and quality control rely on ER-associated degradation (ERAD) and unfolded protein response (UPR). UPR is a feedback control system that acts as a self-protection mechanism in ER stress through detecting unfolded protein pressure in the ER lumen through intracellular signaling, and three pathways are involved, namely ATF6, IRE1, and PERK [19,20]. Note that the ER provides an oxidative environment for protein folding, and protein oxidation increases ROS levels, leading to oxidative stress [21]. It has been demonstrated that prolonged ER stress leads to apoptosis, initiated by the IRE1-XBP1 pathway that enhances the transcriptional expression of the molecular chaperone protein C/EBP CCAAT enhancer binding protein (CHOP), the key factor of apoptosis induced by ER stress, to promote apoptosis [22]. Moreover, ATF6 migrates to the nucleus in response to nuclear localization signals and induces the expression of CHOP [23]. In addition, the disruption of ER calcium homeostasis leads to cytoplasmic Ca^{2+} overload, resulting in the activation of calpain, activation of the mitochondrial pathway, and the caspase cascade triggering apoptosis [24]. ER stress and the pro-apoptotic effects of the UPR pathway contribute largely to kidney injury. The mechanism of ER stress-related pathways induced by AFB1 is presented in Figure 1.

The contamination of animal feed with AFB1 is a significant threat to animal health. Different livestock species have varying susceptibility to AFB1, and donkeys are more sensitive to AFB1 compared with ruminants due to their monogastric nature and their use of the cecum for plant fiber digestion [5]. Therefore, an AFB1 exposure donkey foal model has been established in this study to investigate the nephrotoxic effects of AFB1 on donkey. Kidney function, free radical contents, and antioxidant enzyme activities were detected by corresponding kits. The expression levels of EndoG, ER stress-related, and apoptosis-related genes were tested by RT-qPCR and Western blotting. The objective is to understand the potential mechanisms underlying AFB1-induced nephrotoxicity in donkeys and provide insights for mitigating AFB1 toxicity.

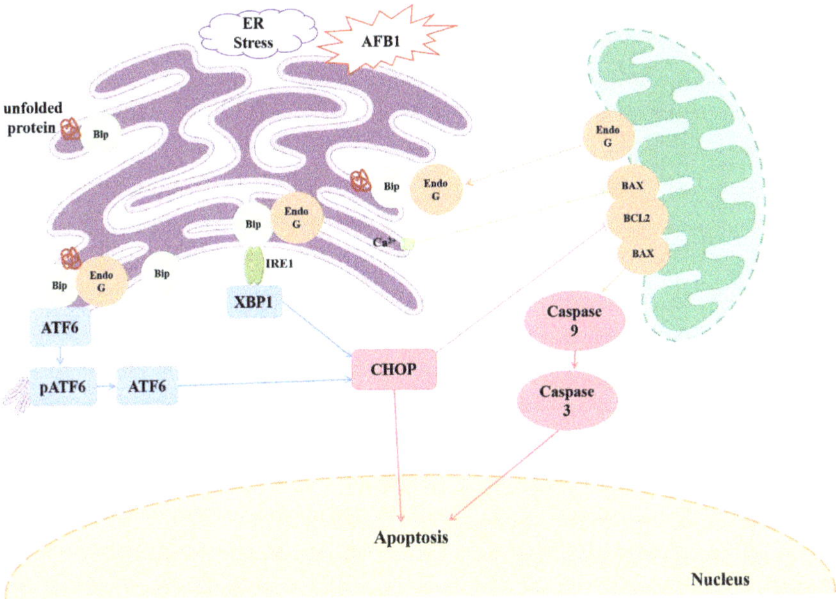

Figure 1. Schematic diagram of the AFB1-induced endoplasmic reticulum stress-related pathway.

2. Materials and Methods

2.1. Animal and Treatment

All procedures were approved by the Animal Welfare and Ethics Committee of the Institute of Animal Science, Liaocheng University (protocol no. 2022112001). Ten 6-month-old weaned male donkey foals (Dezhou donkey, Dezhou, China) with similar weight and body conditions, were randomly and equally divided into two groups, the control group and AFB1-exposed group. The control group was fed a full-price diet (Dong'e Liuhe Lvjia Feed Co., Ltd., Liaocheng, China) and the AFB1-exposed group was fed a full-price diet containing 1 mg AFB1/kg of diet (AFB1, purity $\geq$ 99.9%, Qingdao Prebon Bioengineering Co. Ltd., Qingdao, China). Donkeys were feed a 1 kg diet at 8:00 a.m. and 15:00 p.m., respectively. Moreover, all foals were housed in individual pens in a donkey barn and had free access to forage grass and water during the experiment. After 30 days of feeding, the foals were slaughtered and blood and kidney samples were collected, processed, and stored in accordance with the study requirements.

2.2. Routine Blood and Renal Function Tests

The blood was collected using an EDTA anticoagulant tube and tested in a veterinary automatic blood cell analyzer (BC-5000Vet, Shenzhen Myriad Animal Medical Technology Co., Ltd., Shenzhen, China) to observe the changes in the number and morphological distribution of blood cells.

The serum was obtained from whole blood through centrifugation. It was then added to the BS-180, a fully automatic biochemical analyzer manufactured by Shenzhen Myriad Animal Medical Technology Co., Ltd., China. The analyzer was equipped with test kits specifically designed for measuring the levels of uric acid (UA, D007-a), urea (UREA, D009-a), and creatinine (CREA, D008-a) in the blood.

2.3. Determination of Renal Organ Coefficient

The foal's body weight and the weight of its fresh kidney tissue were measured to calculate the renal organ coefficient of the foal. The formula is as follows:

$$\text{Renal organ coefficient} = \text{kidney weight (kg)}/\text{body weight (kg)}.$$

2.4. Histomorphological Observation

Kidney tissues (1 cm × 1 cm × 0.2 cm) were promptly collected and fixated in a 4% paraformaldehyde buffer (Servicebio, Wuhan Service Biotechnology Co., Wuhan, China) for a minimum of 24 h. The samples underwent the conventional paraffin embedding technique, involving obtaining sections from prepared paraffin blocks. These sections were then degreased with xylene (Chinese national medicine, Beijing Sinopharm Chemical Reagent Co., Ltd., Beijing, China), rehydrated with graded alcohol, and stained with hematoxylin and eosin (Servicebio, Wuhan Service Biotechnology Co., China). The resulting HE-stained sections were examined under a microscope, and images were captured for further analysis.

2.5. Ultrastructure Observations

Kidney samples (1 mm^3) were fixed in a solution of 2.5% glutaraldehyde phosphate sodium buffer (Chinese national medicine, Beijing Sinopharm Chemical Reagent Co., Ltd., China) at a temperature of 4 °C. They were then washed with 0.1 M PBS (pH 7.2, Servicebio, Wuhan Service Biotechnology Co., China) and subsequently treated with osmium tetroxide buffer (Chinese national medicine, Beijing Sinopharm Chemical Reagent Co., Ltd., China) for secondary fixation. To prepare ultrathin sections, a gradient elution using ethanol and acetone (Chinese national medicine, Beijing Sinopharm Chemical Reagent Co., Ltd., China) was performed, followed by soaking and embedding with epoxy. The sections were stained with Mg–uranyl acetate and lead citrate (Chinese national medicine, Beijing Sinopharm Chemical Reagent Co., Ltd., China) and finally examined and imaged using a transmission electron microscope (H-7650, Hitachi Limited, Tokyo, Japan).

2.6. TUNEL Assay

Apoptotic cells in kidneys were detected according to the instructions of TUNEL Apoptosis Detection Kit (Servicebio, Wuhan Service Biotechnology Co., China) using the paraffin section. The scanning pictures were observed and obtained using the imaging system of a fluorescence microscope (Nikon DS-U3, Nippon Kogaku Kogyo Co., Tokyo, Japan). DAPI-stained nuclei were blue under UV excitation, and the nuclei of positive apoptotic cells were green.

2.7. Immunofluorescence Staining

Paraffin sections of kidney were subjected to antigen retrieval after deparaffinization and rehydration. Then, the sections were blocked with 5% BSA–TBSTx (Solarbio, Beijing Suolaibao Techbology Co., Ltd., Beijing, China) for 1 h at room temperature and incubated with the EndoG (1:200) primary antibody at 4 °C overnight. After washing with TBSTx (Solarbio, Beijing Suolaibao Techbology Co., Ltd., China), sections were incubated with TRITC goat anti-rabbit immunoglobulin G (IgG) for 1 h at room temperature. Finally, DAPI (Solarbio, Beijing Suolaibao Techbology Co., Ltd., China) was used to label the nucleus after washing again. The results were observed and pictures were taken using a fluorescence microscope (IX50; Tokyo Olympus Corporation, Tokyo, Japan), which were analyzed by ImageJ, 1.54k.

To test whether EndoG transferred to the ER, EndoG and GRP78 (ER marker, 1:200) were incubated at one section using homologous double-labeled immunostaining. Other processes were consistent with the general immunofluorescence.

2.8. Detection of Antioxidant Function

The contents of total protein (TP, A045-2-2), inducible nitric oxide synthase (iNOS, A014-1), nitric oxide (NO, A012-1), hydrogen peroxide (H_2O_2, A064-1-1), malondialdehyde (MDA, A003-1), and the activity of total antioxidant capacity (T-AOC, A015-1) in donkey kidney were detected by spectrophotometry using the supernatant of the kidney homogenate according to the instruction manual of diagnostic kits (Jiancheng Bioengineering Institute, Nanjing, China). The results were calculated with the formula offered in the instructions based on the OD values at 595 nm (TP), 530 nm (iNOS), 550 nm (NO), 405 nm (H_2O_2), 532 nm (MDA), and 520 nm (T-AOC).

2.9. Real-Time Quantitative PCR Analysis

Total RNA was extracted from kidneys using the classical Trizol method, then total RNA was reverse-transcribed into cDNA using the PrimeScript™ RT reagent Kit with gDNA Eraser (Takara, Dalian Baoji-Medical Technology Co., Ltd., Dalian, China). The cycle threshold (Ct) values of target genes were detected on the CFX96 Touch Real-Time PCR Detection System (Bio-Rad, Bio-Rad Laboratories, Hercules, CA, USA) using TB Green®® Premix Ex Taq™ II (Tli RNaseH Plus) (Takara, Dalian Baoji-Medical Technology Co., Ltd., Kusatsu City, Japan). The reaction system was as follows: TB Green®® Premix Ex Taq™ II, 12.5 μL; forward primer, 1 μL; reverse primer, 1 μL; cDNA, 2 μL; and DEPC H_2O, 8.5 μL. Amplification procedures were as follows: 95 °C for 30 s; 95 °C for 5 s; and 60 °C for 30 s, 39 cycles. Table 1 lists the primers used in this study, and the primer solubilization curves are single peaks. The relative mRNA expression levels of target genes were calculated by the method of $2^{-\Delta\Delta Ct}$ based on the internal reference gene actin (β-actin).

Table 1. Primers used in this study.

Name	Forward Primer Sequence (5′-3′)	Reverse Primer Sequence (5′-3′)
β-actin	CGGGACCTGACGGACTACCTC	TCCTTGATGTCACGCACGATTTCC
Endo G	TCTGCTCCTATGTGATGCCCAAC	CTTGACTCTGCCCGCCCTTG
GRP78	AACCGCATCACGCCGTCTTATG	GGTTGGAGGTGAGCTGGTTCTTG
GRP94	ACCCCGATGCAAAGGTTGAAGAAG	GTCTTGCTCCGTGTCGTCTGTG
ATF-6	TGGGAAACAGGCATTTGGGACATC	CTGAACAACTTGAGGAGGCTGGAG
IRE1	TCCAACCACTCGCTCCACTCTAC	CCTCATCCTCGTCGTCCTGCTC
XBP1	ACTGAAGAGGAGGCTGAGACCAAG	GGAGAGGTTCTGGAGGGGTGAC
CHOP	TGCTTCTCTGGCTTGGCTGAC	TGGTCTTCCTCCTCTTCCTCCTG
BAX	TGGACACTGGACTTCCTTCGAGAG	TGGTGAGCGAGGCGGTGAG
BCL2	GGGACGCTTTGCCACGGTAG	CGGTTGACGCTCTCCACACAC
Caspase9	GACCTGACCGCCGAGCAAATG	TGACAGCCGTGAGAGAGGATGAC
Caspase3	TGCAGAAGTCTAGCTGGAAAACCC	TAGCACAAAGCGACTGGATGAACC

2.10. Western Blotting Analysis

Total proteins were extracted from kidneys by radioimmunoprecipitation assay (RIPA) lysis buffer (Beyotime, Shanghai Biyuntian Biotechnology Co., Ltd., Shanghai, China) with the addition of 1 mM PMSF (Beyotime, Shanghai Biyuntian Biotechnology Co., Ltd., China) and 2% phosphatase inhibitor (Beyotime, Shanghai Biyuntian Biotechnology Co., Ltd., China). The protein concentration was then determined according to the instructions of BCA Protein Concentration Assay Kit (Solarbio, Beijing Solarbio Technology Co., Ltd., China). Equal amounts of total protein were subjected to SDS-PAGE gel electrophoresis and transferred onto a nitrocellulose membrane (Biosharp, Beijing Lanjieke Science and Technology Co., Ltd., Beijing, China). The membranes were blocked in 5% TBST–skim milk for 1 h at 37 °C and then incubated with the primary antibody at 4 °C overnight. After washing three times, the membranes were incubated with HRP goat anti-rabbit IgG (1:10,000, Abclonal, Wuhan ABclonal Technology Co., Ltd., Wuhan, China) at room temperature for 120 min. The positive signal was detected on the chemiluminescence system using an enhanced chemiluminescence kit (Meilunbio, Dalian Meilun Biotechnology Co., Ltd., Dalian, China). Images were collected by the ChemiDocTM XRS+ imaging system and then analyzed by Image J. β-actin was used as an internal reference. The primary antibodies used in this study are listed in Table 2.

Table 2. Brands and dilutions of primary antibodies.

Antibody Name	Diluting Factor	Source	Brand
β-actin	1:10,000	Rabbit	Abclonal
EndoG	1:1000	Rabbit	Abclonal
GRP78	1:2000	Rabbit	Wanleibio
GRP94	1:1000	Rabbit	Wanleibio
ATF6	1:2000	Rabbit	Wanleibio
IRE1	1:1000	Rabbit	Wanleibio
XBP1	1:1000	Rabbit	Wanleibio
BAX	1:1000	Rabbit	Wanleibio
BCL2	1:1000	Rabbit	Wanleibio
Caspase9	1:1000	Rabbit	Wanleibio
Caspase3	1:1000	Rabbit	Wanleibio

Note: Abclonal, Wuhan ABclonal Technology Co., Ltd., China; Wanleibio, Shenyang Wanbei Biotechnology Co., Shenyang, China.

2.11. Data and Analyses

The experimental data were statistically analyzed using WPS office and graphically analyzed using GraphPad Prism 8.0.2. Data are expressed as mean ± SEM ($n = 5$), and the differences between the two groups were analyzed using the *t*-test. "*" denotes a significant difference compared to the control group ($p < 0.05$); "ns" indicates no significant difference compared to the control group ($p > 0.05$).

3. Results

3.1. Effect of AFB1 on Blood Indices in Donkey

As shown in Table 3, compared to the control group, the AFB1 group did not show any significant effects on white blood cell counts (WBC), hemoglobin (HGB), platelets (PLTs), red blood cell counts (RBC), and lymphocytes (Lym). However, there was a significant increase in neutrophil number (Neu), neutrophil percentage (Neu %), and lymphocyte percentage (Lym %) in the AFB1 group ($p < 0.05$). These findings suggest that AFB1 exposure leads to inflammation in donkeys, disrupting their homeostasis and causing damage.

Table 3. Routine testing of donkey blood.

	Control Group	AFB1 Group	*p*-Value
WBC (10^9/L)	14.73 ± 1.23	13.94 ± 0.88	0.27
HGB (g/L)	117.2 ± 9.83	105.6 ± 10.6	0.11
PLT (10^9/L)	301 ± 72.6	328.25 ± 36.3	0.53
RBC (10^{12}/L)	6.88 ± 0.51	6.22 ± 0.72	0.133
Lym (10^9/L)	6.82 ± 1.85	7.8 ± 1.92	0.194
Neu (10^9/L)	4.91 ± 0.29	6.52 ± 0.78	0.002
Neu%	35.3 ± 3.17	44.62 ± 7.38	0.03
Lym%	46.14 ± 6.14	55.82 ± 3.67	0.015

Note: Values for control and AFB1 are mean ± standard deviation, and *p*-values were derived from *t*-tests.

3.2. Impact of AFB1 on Renal Organ Coefficient and Renal Function in Donkeys

As shown in Figure 2A, the renal organ coefficient in the AFB1 group was significantly higher compared to the control group ($p < 0.05$). Additionally, the levels of UA, CREA, and UREA in the serum were significantly elevated in the AFB1 group ($p < 0.05$) (Figure 2B–D). These findings suggest that AFB1 exposure could result in kidney enlargement and impair kidney function in donkeys.

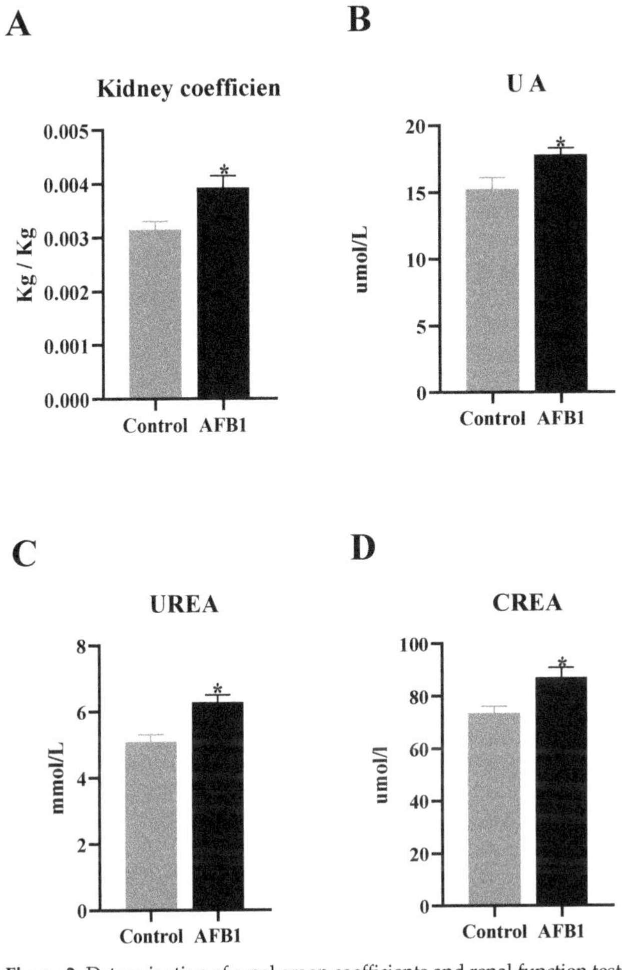

Figure 2. Determination of renal organ coefficients and renal function tests in donkeys. (**A**) Kidney organ coefficient; (**B**) UA; (**C**) UREA; (**D**) CREA, "*" denotes a significant difference compared to the control group ($p < 0.05$).

3.3. Effect of AFB1 on Histopathology of Donkey Kidney Tissue

The kidney tissues in the control group displayed normal morphologies. The glomerulus and its surrounding renal tubules were intact and not detached. The lumen of the balloon was clearly visible (Figure 3A). However, the kidney tubular epithelial cells were congested and detached in the AFB1 group (Figure 3B, red frame). The glomerulus was also congested (Figure 3B, black arrows), and there was a reduction in the gap between the glomerulus and renal capsule (Figure 3B, red arrow). Additionally, casts appeared in the renal tubules (Figure 3C, green arrow). Furthermore, the TUNEL staining results shown that AFB1 exposure increased the number of apoptotic cells compared to the control group (Figure 3D). These findings demonstrate that AFB1 caused pathological damage and apoptosis in the donkey kidney.

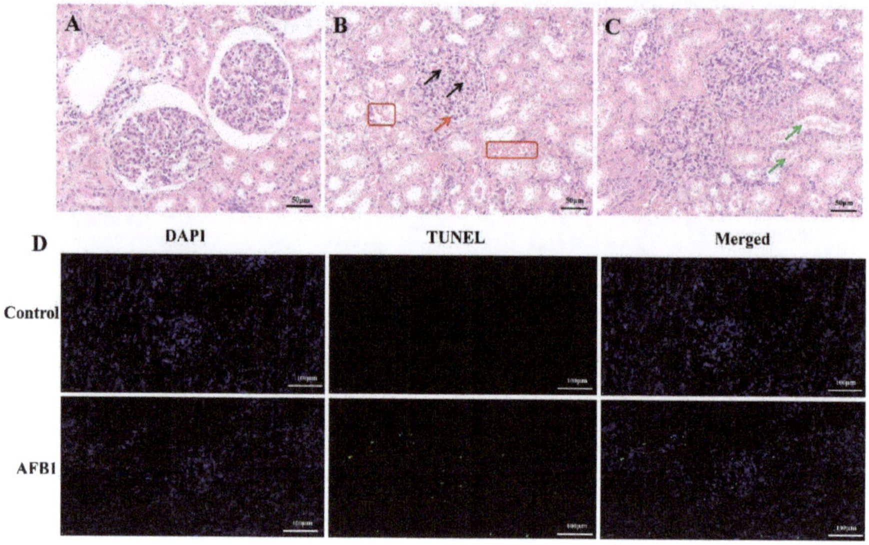

Figure 3. Kidney histopathological observation in donkey. (**A–C**) HE staining results ((**A**) control group; (**B**,**C**) AFB1 group); (**D**) TUNEL staining results. The original images in the manuscript are published as Supplementary Materials.

3.4. *Effect of AFB1 on Ultrastructure of Kidney in Donkey*

As shown in Figure 4A, the control group did not exhibit any obvious abnormalities. However, the AFB1 group, when compared to the control group, displayed typical apoptotic features in the ultrastructure of the kidney. These features included nuclear wrinkling, nuclear deformation, and chromatin aggregation (Figure 4B, red arrow). Additionally, there was evidence of mitochondrial aggregation, swelling, disappearance of crests, and vacuolization (Figure 4B, orange arrows), as well as the accumulation of lipid droplets (Figure 4B, yellow arrows) and the dissolution and fragmentation of the ER (Figure 4B, blue arrows). These findings suggest that exposure to AFB1 can induce apoptosis in the kidney, with ER and mitochondrial damage likely playing significant roles.

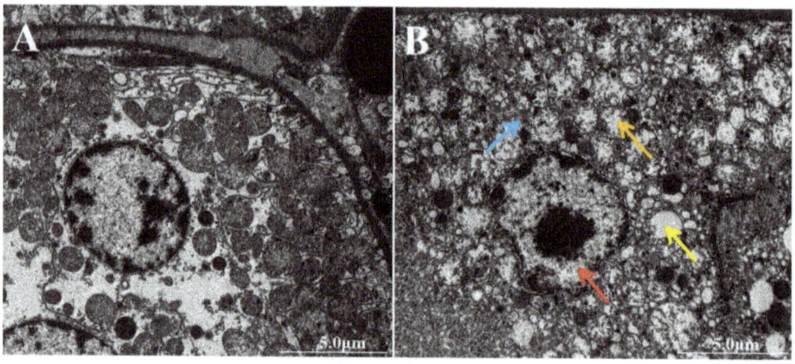

Figure 4. The ultrastructure observation of kidney in donkey. (**A**) Control group; (**B**) AFB1 group.

3.5. Effect of AFB1 on Antioxidant Capacity in Kidney of Donkey

As shown in Figure 5, the AFB1-exposure-group donkeys showed significantly higher levels of iNOS (Figure 5A), NO (Figure 5B), H_2O_2 (Figure 5C), MDA (Figure 5D), and T-AOC (Figure 5E) in kidney compared with the control group ($p < 0.05$). The results indicate that AFB1 exposure causes oxidative stress to the kidneys of donkeys and a compensatory increase in antioxidant capacity to cope with oxidative stress.

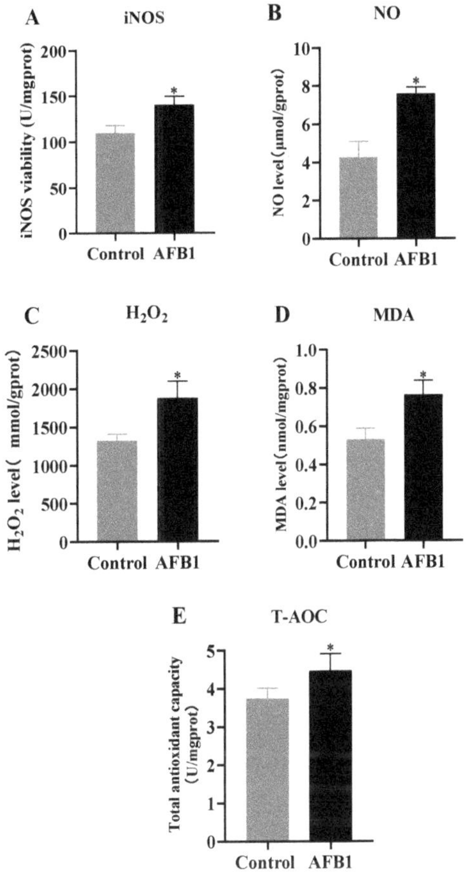

Figure 5. Levels of antioxidant capacity of donkey kidney. (**A**) iNOS; (**B**) NO; (**C**) H_2O_2; (**D**) MDA; (**E**) T-AOC, "*" denotes a significant difference compared to the control group ($p < 0.05$).

3.6. Effect of AFB1 on the EndoG Expression in Donkey Kidney

As shown in Figure 6, the immunofluorescence staining (Figure 6A), mRNA expression (Figure 6B), and protein expression (Figure 6C) results of EndoG confirmed that AFB1 exposure increased the expression of EndoG ($p < 0.05$). Moreover, the immunofluorescence double-staining results of EndoG and GRP78 show that AFB1 exposure increases the expression of GRP78 and the colocalization of EndoG and GRP78 (ER) (Figure 6E). These results suggest that AFB1 exposure induces EndoG expression and promotes EndoG translocation to the ER and binds to GRP78 in donkey kidneys.

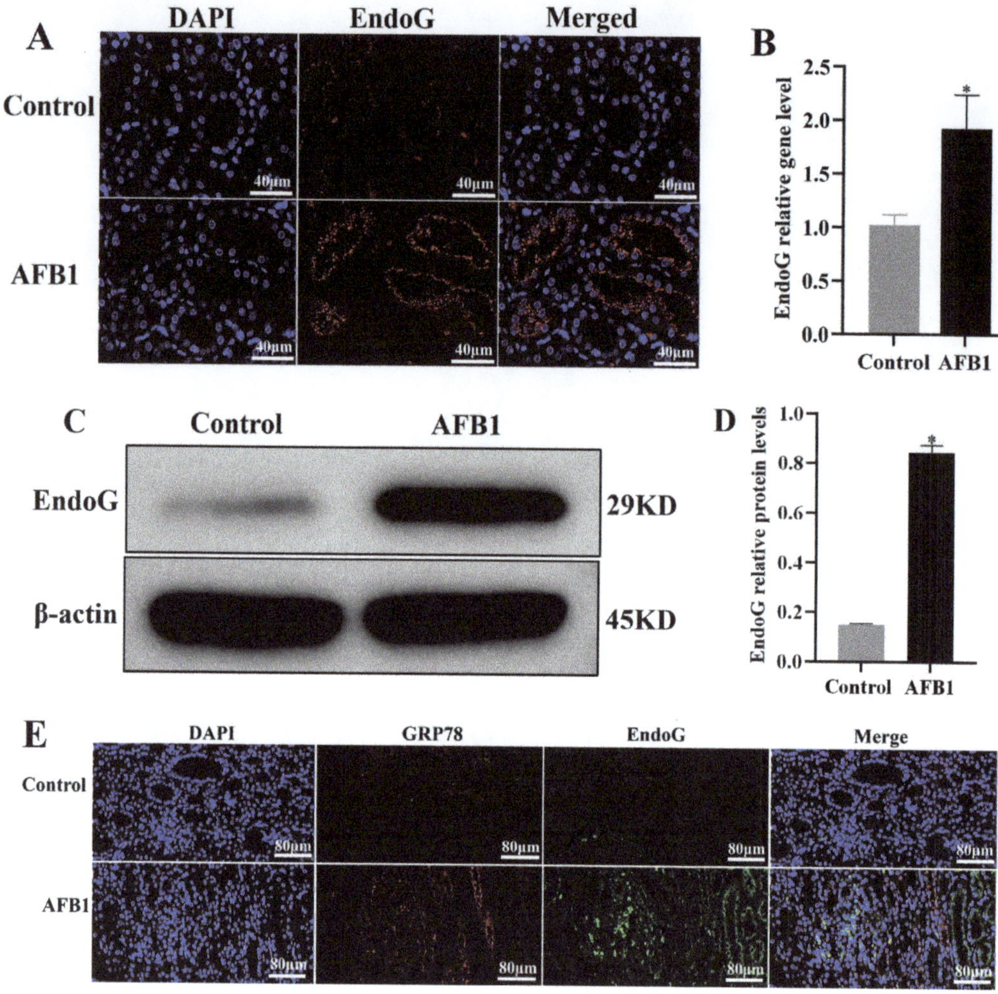

Figure 6. EndoG expression detection in donkey kidney. (**A**) EndoG immunofluorescence results; (**B**) mRNA expression; (**C**) Western blot results; (**D**) protein expression; (**E**) co-immunofluorescence staining of GRP78 and EndoG, where red represents GRP78, green represents EndoG, and blue represents nuclei, "*" denotes a significant difference compared to the control group ($p < 0.05$).

3.7. Effects of AFB1 Exposure on ER Stress in Donkey Kidney

As shown in Figure 7, the ER stress marker GRP78 was significantly elevated at both the gene and protein levels ($p < 0.05$, Figure 7A,B,F). GRP94 was significantly expressed at the gene level ($p < 0.05$, Figure 7A) but not at the protein level ($p > 0.05$, Figure 7C,F). The expressions of ATF-6, IRE1, XBP1, and CHOP were significantly elevated at gene and protein levels ($p < 0.05$, Figure 7A–F). These results indicate that AFB1 exposure induces ER stress in the donkey kidney, and it might promote the expression of CHOP through activating the IRE1/ATF6 pathway.

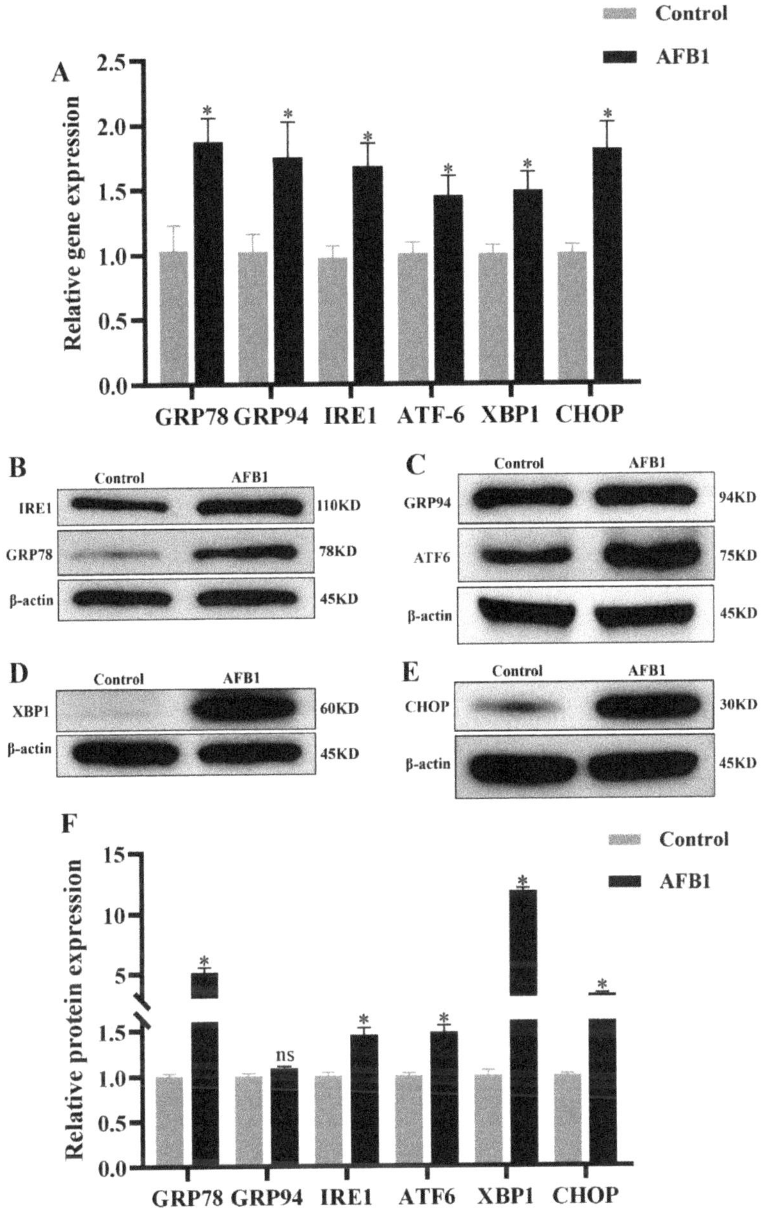

Figure 7. ER stress-related mRNA and protein expression. (**A**) mRNA expression; (**B–E**) Western blot results; (**F**) protein expression, "*" denotes a significant difference compared to the control group ($p < 0.05$); "ns" in-dicates no significant difference compared to the control group ($p > 0.05$).

3.8. Effects of AFB1 Exposure on Apoptosis in Donkey Kidney

As shown in Figure 8, the mRNA and protein expression levels of BAX (Figure 8A,B,F), caspase9 (Figure 8A,D,F), and caspase3 (Figure 8A,E,F) in the AFB1 group were significantly higher than the control group ($p < 0.05$). Conversely, BCL2 exhibited a significantly lower expression ($p < 0.05$, Figure 8A,C,F). These results indicated that the AFB1 exposure activated the mitochondrial apoptotic pathways.

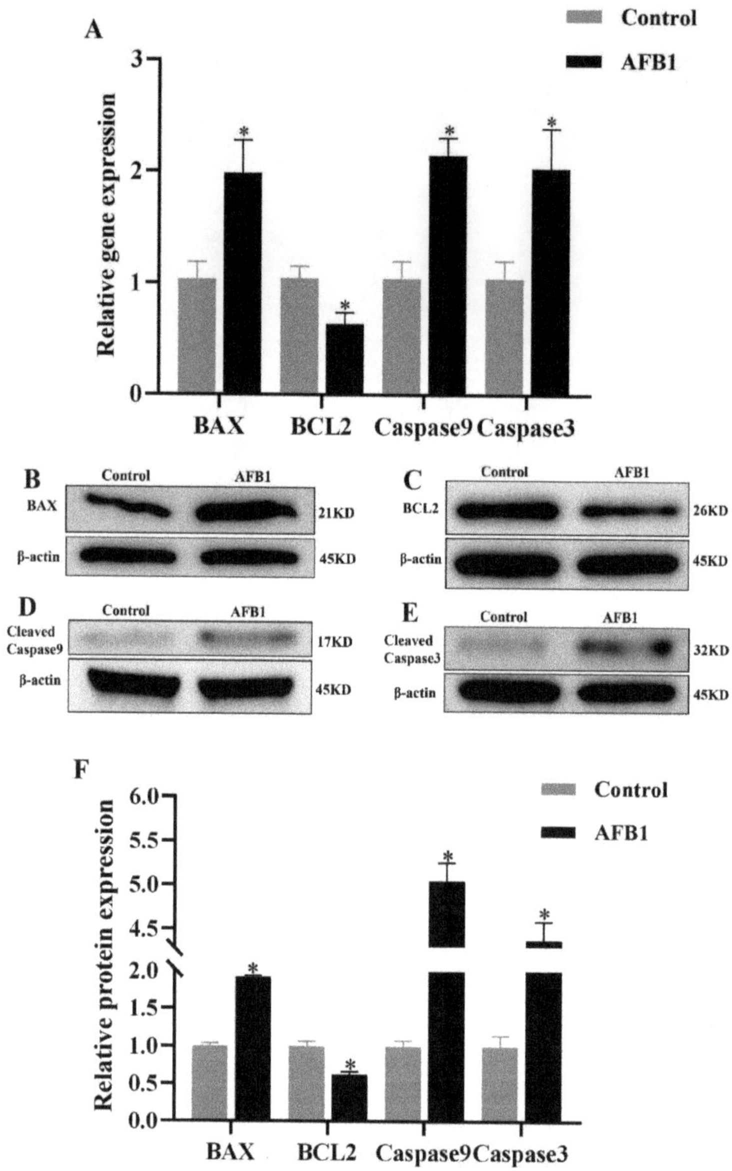

Figure 8. Apoptosis-related mRNA and protein expression. (**A**) mRNA expression; (**B–E**) Western blot results; (**F**) protein expression, "*" denotes a significant difference compared to the control group ($p < 0.05$).

4. Discussion

AFB1 is one of common contaminants in animal feed and forage grasses; AFB1 exposure could cause damage to animal organisms such as hepatotoxicity, nephrotoxicity, and neurotoxicity, and it poses a potential health risk to humans [25–31]. In this study, we first demonstrated the partial mechanism by which AFB1 causes kidney injury in donkey. These data from the AFB1-exposure donkey foal model confirmed that prolonged exposure to AFB1 led to kidney damage in donkey, including swelling and hemorrhage, subcellular organelle injury, and oxidative stress. Furthermore, AFB1 exposure caused a high expression of EndoG, translocation to the ER, and binding to GRP78, and it also induced ER stress (ATF6 and IRE1 pathway), which eventually activated the mitochondrial apoptosis pathway.

It is widely acknowledged that AFB1 exposure affects various animals and humans in terms of health condition and life safety. AFB1 exposure can impair digestion in pigs, break down the intestinal barrier, and inhibit their growth [11]. Cattle exposed to AFB1 exhibit behavioral changes, such as depression and anorexia, and organ changes such as liver enlargement, gallbladder enlargement, and congestion of the intestines and kidneys [12]. A routine blood test is a very convenient and fast inspection, and it is also necessary for basic inspections to determine whether the body is in an abnormal condition. It was found that Neu, Neu %, and Lym % were increased by AFB1 exposure, which indicated that AFB1 exposure caused inflammation in donkeys, disrupted their homeostasis, and induced damage. This result was similar to the research of Oa Mogilnaya et al. in mice in contrast to the study by Raúl Bodas et al. in cows [32,33]. This difference may be attributed to species differences or the duration of AFB1 exposure. Kidney injury was considered as one characteristic of AFB1 exposure. UA, CREA, and UREA are important indexes for kidney function, whose contents commonly can increase under kidney impairment caused by various toxic substance exposures. Chromium or zearalenone exposure causes kidney damage and elevated UA and CREA in mice [34,35]. AFB1 exposure similarly resulted in a significant elevation of UA and CREA in mice and chickens [36,37]. In this study, the contents of UA, CREA, and UREA were increased in AFB1-exposed donkeys, which suggested that AFB1 exposure could cause kidney damage in donkey as well as the animals mentioned above.

Oxidative stress is the key mechanism of AFB1 toxicity. AFB1 disturbs the activities of antioxidant enzymes, leading to peroxide accumulation and excessive ROS-induced damage to proteins and nucleic acids, resulting in the production of large amounts of MDA. AFB1 exposure induces oxidative stress by causing a significant elevation of MDA and T-AOC in the testes of sheep [38]. AFB1 exposure was able to cause a significant elevation of H_2O_2, MDA, and T-AOC in the kidney tissues of mice and a significant elevation of MDA and T-AOC in the kidney tissues of rabbits, causing oxidative stress [39,40]. We found the iNOS activity, H_2O_2, NO, and MDA contents were increased in AFB1-exposure donkey kidneys, which indicated that AFB1 exposure induced oxidative stress, causing kidney oxidative damage. Moreover, the T-AOC was also increased in AFB1-exposure donkey kidneys, which may be a compensatory increase against oxidative damage. Oxidative stress can cause damage to organelles, especially mitochondria [41]. As the centers of intracellular redox reactions, mitochondria are the main targets of oxidative stress, and long-term oxidative stress may lead to mitochondrial damage and dysfunction [42,43].

EndoG, an endonuclease found in the inner membrane gap of mitochondria, is released into the cytoplasm when mitochondrial homeostasis is disrupted by an external stimulus [44]. Furthermore, EndoG in the cytoplasm can further enter the ER under conditions of a high-fat diet and ER stress inducers, binding to Bip to elicit an ER stress response [17]. It was shown that GRP78 and GRP94 were significantly elevated in chicken bursa and thymus by AFB1 exposure, and GRP78, IRE1, XBP1, and CHOP were significantly elevated in mouse hippocampal neurons; thus, exposure to AFB1 resulted in ER stress [45–47]. In this study, AFB1 exposure upregulated the expression of EndoG and GRP78 and led to EndoG accumulation in the ER, identifying AFB1 exposure-induced ER stress in donkey kidney. ER transmembrane sensors (PERK, IRE1, and ATF6) can relieve ER stress and restore ER homeostasis by reducing the accumulation of ER proteins [48]. However, continuing ER stress triggers the activation of the apoptotic pathway, ultimately resulting in cell apoptosis [49]. CHOP is an important molecule involved in the transition of ER stress from anti-apoptosis to pro-apoptosis [50]. The activation of PERK, XBP1, and ATF6 can mediate the production of CHOP, which promotes apoptosis [51,52]. In the present study, AFB1 exposure activated the IRE1/ATF6-CHOP pathway, which caused endoplasmic reticulum stress in renal tissues.

Mitochondria play a crucial role in cell apoptosis. Mitochondria receive apoptotic signals released by external stimuli, which lead to mitochondrial outer membrane permeability (MOMP) and trigger the release of a series of proteins within the mitochondria, inducing the mitochondrial apoptosis pathway [37]. The BCL2 family plays key roles in regulating mitochondrial outer membrane integrity and function. CHOP plays a crucial role in this process due to its regulation on the BCL2 family of proteins, specifically down-regulating BCL2 expression and increasing the expression of BAX [53,54]. As a pore-forming protein, BAX transfers from the cytoplasm to the mitochondria, leading to mitochondrial outer membrane permeability (MOMP) and triggering apoptosis. The inhibition of BCL2, the transference of BAX to mitochondria, the release of cytochrome C into the cytosol, and the activation of apoptotic initiators and executioners (for caspases) primarily involve several key events of the mitochondrial apoptotic pathway [55–57]. Studies have shown that exposure to AFB1 increases BAX and caspase3 levels, decreases BCL2 levels, activates the mitochondrial apoptotic pathway, and leads to apoptosis in mouse testes and chicken liver [58,59]. The Tunel and ultrastructure observation results showed that AFB1 induced donkey kidney cell apoptosis. Additionally, the decrease in BCL2 expression and the increase in BAX, caspase9, and caspase3 expression suggested that AFB1 exposure activated the mitochondrial apoptosis pathway.

5. Conclusions

Altogether, our study concluded that the donkey kidney exhibits notable sensitivity to AFB1, resulting in pathological changes. Furthermore, AFB1 exposure promotes the production of ROS, leading to oxidative stress in the donkey kidney. In response, a compensatory increase in T-AOC was observed. Mechanistic investigations further revealed that AFB1 exposure activated EndoG, which triggered ER stress and subsequently activated the IRE1/ATF6-CHOP signaling pathway. This cascade ultimately induced mitochondrial apoptosis. As a result, AFB1 exposure caused significant kidney damage and dysfunction. Overall, our findings contribute to a deeper understanding of the mechanisms underlying AFB1-induced nephrotoxicity in donkeys. Importantly, these results provide a foundation for future research aimed at mitigating AFB1 toxicity, such as strategies to alleviate oxidative stress through ROS removal or the enhancement of antioxidant defenses, as well as approaches to inhibit EndoG expression and alleviate ER stress to reduce kidney injury.

Supplementary Materials: The following are available online at https://www.mdpi.com/article/10.3390/vetsci12020130/s1, The original images in the manuscript are published in the Supplementary Materials.

Author Contributions: Conceptualization, Y.J., Y.Z., W.S., X.Z., and W.L.; data curation, Y.J. and W.S.; formal analysis, Y.J.; investigation, Y.J., Y.Z., J.G., and W.L.; methodology, Y.J., Y.Z., W.S., G.L., M.Z.K., X.Z., and W.L.; project administration, X.Z. and W.L.; resources, G.L., C.W., X.Z., and W.L.; software, Y.J., Y.Z., W.S., and X.Z.; supervision, X.Z. and W.L.; validation, W.S., M.Z.K., and W.L.; visualization, Y.J., J.G., G.L., C.W., M.Z.K., and W.L.; writing—original draft, Y.J., X.Z., and W.L.; writing—review and editing, Y.J., W.S., J.G., G.L., C.W., M.Z.K., X.Z., and W.L. All authors have read and agreed to the published version of the manuscript.

Funding: This research was funded by the Natural Science Foundation of Shandong Province (ZR2023MC209), the Shandong Donkey Industry Technology System Fund (SDAIT-27-11), and the Liaocheng University Science and Technology Tackling Project (318011701, 318052242).

Institutional Review Board Statement: All procedures were approved by the Animal Welfare and Ethics Committee of the Institute of Animal Science, Liaocheng University (protocol no. 2022112001).

Informed Consent Statement: Informed consent was obtained from all subjects involved in the study.

Data Availability Statement: All the data are available in this article.

Acknowledgments: Special thanks go to Dong'e e jiao corporation limited for providing help and support during the experiment.

Conflicts of Interest: The authors declare that they have no known competing financial interests or personal relationships that could have appeared to influence the work reported in this paper.

References

1. Gurtoo, H.L.; Dahms, R.P.; Paigen, B. Metabolic Activation of Aflatoxins Related to Their Mutagenicity. *Biochem. Biophys. Res. Commun.* **1978**, *81*, 965–972. [CrossRef] [PubMed]
2. Bondy, G.S.; Pestka, J.J. Immunomodulation by Fungal Toxins. *J. Toxicol. Pestka Environ. Health Part B* **2000**, *3*, 109–143.
3. Abdel-Fattah, H.M.; Kamel, Y.Y.; Megalla, S.E.; Mycopathologia, H.A.H. Aflatoxin and Aflatoxicosis. *Mycopathologia* **1982**, *77*, 129–135. [CrossRef] [PubMed]
4. Rodricks, J.V.; Hesseltine, C.W.; Mehlman, M.A. Mycotoxins in Human and Animal Health. In Proceedings of the Conference on Mycotoxins in Human and Animal Health, Convened at University of Maryland University College, Center of Adult Education, College Park, MD, USA, 4–8 October 1977.
5. Zou, G.X.; Zhang, H.X.; Hua, R.M. Research Progress in Toxicological Effects and Mechanism of T-2 Toxin. *Asian J. Ecotoxicol.* **2011**, *6*, 121–128.
6. Bertrand, G.; Applegate, T. Modulation of Intestinal Functions Following Mycotoxin Ingestion: Meta-Analysis of Published Experiments in Animals. *Toxins* **2013**, *5*, 396–430. [CrossRef]
7. Peters, A.; Nawrot, T.S.; Baccarelli, A.A. Hallmarks of Environmental Insults. *Cell* **2021**, *184*, 1455–1468. [CrossRef]
8. Benkerroum, N. Chronic and Acute Toxicities of Aflatoxins: Mechanisms of Action. *Int. J. Environ. Res. Public Health* **2020**, *17*, 423. [CrossRef] [PubMed]
9. Cao, W.Y.; Yu, P.; Yang, K.P.; Cao, D.L. Aflatoxin B1: Metabolism, Toxicology, and Its Involvement in Oxidative Stress and Cancer Development. *Toxicol. Mech. Methods* **2022**, *32*, 395–419. [CrossRef] [PubMed]
10. Deng, J.; Zhao, L.; Zhang, N.; Karrow, A.K.; Krumm, Q.D.; Sun, L. Aflatoxin B-1 Metabolism: Regulation by Phase I and Ii Metabolizing Enzymes and Chemoprotective Agents. Mutation Research. *Rev. Mutat. Res.* **2018**, *778*, 79–89. [CrossRef] [PubMed]
11. Pu, J.; Yuan, Q.; Yan, H.; Tian, G.; Chen, D.; He, J.; Zheng, P.; Yu, J.; Mao, X.; Huang, Z. Effects of Chronic Exposure to Low Levels of Dietary Aflatoxin B1 on Growth Performance, Apparent Total Tract Digestibility and Intestinal Health in Pigs. *Animals* **2021**, *11*, 336. [CrossRef]
12. Elgioushy, M.M.; Elgaml, S.A.; El-Adl, M.M.; Hegazy, A.M.; Hashish, E.A. Aflatoxicosis in Cattle: Clinical Findings and Biochemical Alterations. *Environ. Sci. Hashish Pollut. Res. Int.* **2020**, *28*, 35526–35534. [CrossRef] [PubMed]
13. Fouad, A.M.; Ruan, D.; El-Senousey, H.K.; Chen, W.; Jiang, S.; Zheng, C. Harmful Effects and Control Strategies of Aflatoxin B1 Produced by Aspergillus Flavus and Aspergillus Parasiticus Strains on Poultry: Review. *Toxins* **2019**, *11*, 176. [CrossRef] [PubMed]
14. Wang, H.; Li, W.; Muhammad, I.; Sun, X.; Cui, X.; Chen, P.; Qayum, A.; Zhang, X. Biochemical Basis for the Age-Related Sensitivity of Broilers to Aflatoxin B1. *Toxicol Mech Methods* **2018**, *28*, 361–368. [CrossRef] [PubMed]
15. Li, H.; Li, S.; Yang, H.; Wang, Y.; Wang, J.; Zheng, N. L-Proline Alleviates Kidney Injury Caused by Afb1 and Afm1 through Regulating Excessive Apoptosis of Kidney Cells. *Toxins* **2019**, *11*, 226. [CrossRef]
16. Qian, G.; Tang, L.; Lin, S.; Xue, K.S.; Nicole, J.M.; Su, J.; Gelderblom, W.C.; Riley, R.T.; Phillipset, T.D.; Wang, J. Sequential Dietary Exposure to Aflatoxin B1 and Fumonisin B1 in F344 Rats Increases Liver Preneoplastic Changes Indicative of a Synergistic Interaction. *Food Chem. Toxicol.* **2016**, *95*, 188–195. [CrossRef] [PubMed]
17. Wang, W.; Tan, J.; Liu, X.; Guo, W.; Li, M.; Liu, X.; Liu, Y.; Dai, W.; Hu, L.; Wang, Y.; et al. Cytoplasmic Endonuclease G promotes nonalcoholic fatty liver disease via mTORC2-AKT-ACLY and endoplasmic reticulum stress. *Nat. Commun.* **2023**, *14*, 6201. [CrossRef] [PubMed]
18. Berridge, M.J. The endoplasmic reticulum: A multifunctional signaling organelle. *Cell Calcium.* **2002**, *32*, 235–249. [CrossRef] [PubMed]

19. Arruda, A.P.; Hotamisligil, G.S. Calcium Homeostasis and Organelle Function in the Pathogenesis of Obesity and Diabetes. *Cell Metab.* **2015**, *22*, 381–397. [CrossRef] [PubMed]
20. Krebs, J.; Agellon, L.B.; Michalak, M. Ca(2+) homeostasis and endoplasmic reticulum (ER) stress: An integrated view of calcium signaling. *Biochem. Biophys. Res. Commun.* **2015**, *460*, 114–121. [CrossRef] [PubMed]
21. Hourihan, J.M.; Moronetti, M.L.E.; Fernández-Cárdenas, L.P.; Blackwell, T.K. Cysteine Sulfenylation Directs IRE-1 to Activate the SKN-1/Nrf2 Antioxidant Response. *Mol. Cell.* **2016**, *63*, 553–566. [CrossRef]
22. Fernández, A.; Ordóñez, R.; Reiter, R.J.; González-Gallego, J.; Mauriz, J.L. Melatonin and endoplasmic reticulum stress: Relation to autophagy and apoptosis. *J. Pineal. Res.* **2015**, *59*, 292–307. [CrossRef] [PubMed]
23. Zhang, J.; Guo, J.; Yang, N.; Huang, Y.; Hu, T.; Rao, C. Endoplasmic reticulum stress-mediated cell death in liver injury. *Cell Death Dis.* **2022**, *13*, 1051. [CrossRef] [PubMed]
24. Gorman, A.M.; Healy, S.J.; Jäger, R.; Samali, A. Stress management at the ER: Regulators of ER stress-induced apoptosis. *Pharmacol. Ther.* **2012**, *134*, 306–316. [CrossRef] [PubMed]
25. Corcuera, L.A.; Ibáñez-Vea, M.; Vettorazzi, A.; González-Peñas, E.; Cerain, A.L. Validation of a UHPLC-FLD analytical method for the simultaneous quantification of aflatoxin B1 and ochratoxin a in rat plasma, liver and kidney. *J. Chromatogr. B Analyt. Technol. Biomed. Life Sci.* **2011**, *879*, 2733–2740. [CrossRef] [PubMed]
26. Monson, M.S.; Settlage, R.E.; McMahon, K.; Mendoza, K.M.; Rawal, S.; El-Nezami, H.S.; Coulombe, R.A. Reed KM. Response of the hepatic transcriptome to aflatoxin B1 in domestic turkey (Meleagris gallopavo). *PLoS ONE* **2014**, *9*, e100930. [CrossRef] [PubMed]
27. Mykkänen, H.; Zhu, H.; Salminen, E.; Juvonen, R.O.; Ling, W.; Ma, J.; Polychronaki, N.; Kemiläinen, H.; Mykkänen, O.; Salminen, S.; et al. Fecal and urinary excretion of aflatoxin B1 metabolites (AFQ1, AFM1 and AFB-N7-guanine) in young Chinese males. *Int. J. Cancer* **2005**, *115*, 879–884. [CrossRef] [PubMed]
28. Narkwa, P.W.; Blackbourn, D.J.; Mutocheluh, M. Aflatoxin B_1 inhibits the type 1 interferon response pathway via STAT1 suggesting another mechanism of hepatocellular carcinoma. *Infect. Agent. Cancer* **2017**, *12*, 17. [CrossRef]
29. Pandey, I.; Chauhan, S.S. Studies on production performance and toxin residues in tissues and eggs of layer chickens fed on diets with various concentrations of aflatoxin AFB1. *Br. Poult. Sci.* **2007**, *48*, 713–723. [CrossRef]
30. Câmara, N.O.; Iseki, K.; Kramer, H.; Liu, Z.H.; Sharma, K. Kidney disease and obesity: Epidemiology, mechanisms and treatment. *Nat. Rev. Nephrol.* **2017**, *13*, 181–190. [CrossRef] [PubMed]
31. Romagnoli, S.; Ricci, Z. The Kidney in Diastolic Dysfunction. In *Critical Care Nephrology*, 3rd ed.; Elsevier: Amsterdam, The Netherlands, 2019; pp. 718–721.
32. Mogilnaya, O.; Puzyr, A.; Baron, A.; Bondar, V. Hematological parameters and the state of liver cells of rats after oral administration of aflatoxin b1 alone and together with nanodiamonds. *Nanoscale Res. Lett.* **2010**, *5*, 908–912. [CrossRef] [PubMed]
33. Bodas, R.; Giráldez, F.J.; Olmedo, S.; Herrera, M.; Lorán, S.; Ariño, A.; López, S.; Benito, A.; Juan, T. The Effects of Aflatoxin B1 Intake in Assaf Dairy Ewes on Aflatoxin M1 Excretion, Milk Yield, Haematology and Biochemical Profile. *Animals* **2023**, *13*, 436. [CrossRef]
34. Soudani, N.; Sefi, M.; Ben, A.I.; Boudawara, T.; Zeghal, N. Protective effects of Selenium (Se) on Chromium (VI) induced nephrotoxicity in adult rats. *Ecotoxicol. Environ. Saf.* **2010**, *73*, 671–678. [CrossRef]
35. Wang, N.; Li, P.; Wang, M.; Chen, S.; Huang, S.; Long, M.; Yang, S.; He, J. The Protective Role of *Bacillus velezensis* A2 on the Biochemical and Hepatic Toxicity of Zearalenone in Mice. *Toxins* **2018**, *10*, 449. [CrossRef] [PubMed]
36. Gao, X.; Jiang, L.; Xu, J.; Liu, W.; Li, S.; Huang, W.; Zhao, H.; Yang, Z.; Yu, X.; Wei, Z. Aflatoxin B1-activated heterophil extracellular traps result in the immunotoxicity to liver and kidney in chickens. *Dev. Comp. Immunol.* **2022**, *128*, 104325. [CrossRef] [PubMed]
37. Li, W.; Wu, G.; Yang, X.; Yang, J.; Hu, J. Taurine Prevents AFB1-Induced Renal Injury by Inhibiting Oxidative Stress and Apoptosis. *Adv. Exp. Med. Biol.* **2022**, *1370*, 435–444. [PubMed]
38. Lin, L.X.; Cao, Q.Q.; Zhang, C.D.; Xu, T.T.; Yue, K.; Li, Q.; Liu, F.; Wang, X.; Dong, H.J.; Huang, S.C.; et al. Aflatoxin B1 causes oxidative stress and apoptosis in sheep testes associated with disrupting rumen microbiota. *Ecotoxicol. Environ. Saf.* **2022**, *232*, 113225. [CrossRef]
39. Wang, Y.; Liu, F.; Zhou, X.; Liu, M.; Zang, H.; Liu, X.; Shan, A.; Feng, X. Alleviation of Oral Exposure to Aflatoxin B1-Induced Renal Dysfunction, Oxidative Stress, and Cell Apoptosis in Mice Kidney by Curcumin. *Antioxidants* **2022**, *11*, 1082. [CrossRef] [PubMed]
40. Liu, Y.; Wang, J.; Chang, Z.; Li, S.; Zhang, Z.; Liu, S.; Wang, S.; Wei, L.; Lv, Q.; Ding, K.; et al. SeMet alleviates AFB_1-induced oxidative stress and apoptosis in rabbit kidney by regulating Nrf2//Keap1/NQO1 and PI3K/AKT signaling pathways. *Ecotoxicol. Environ. Saf.* **2024**, *269*, 115742. [CrossRef] [PubMed]
41. Xu, F.; Wang, P.; Yao, Q.; Shao, B.; Yu, H.; Yu, K.; Li, Y. Lycopene alleviates AFB_1-induced immunosuppression by inhibiting oxidative stress and apoptosis in the spleen of mice. *Food Funct.* **2019**, *10*, 3868–3879. [CrossRef]

42. Liu, H.; He, Y.; Gao, X.; Li, T.; Qiao, B.; Tang, L.; Lan, J.; Su, Q.; Ruan, Z.; Tang, Z.; et al. Curcumin alleviates AFB1-induced nephrotoxicity in ducks: Regulating mitochondrial oxidative stress, ferritinophagy, and ferroptosis. *Mycotoxin Res.* **2023**, *39*, 437–451. [CrossRef]
43. Tao, W.; Li, Z.; Nabi, F.; Hu, Y.; Hu, Z.; Liu, J. *Penthorum chinense* Pursh Compound Ameliorates AFB1-Induced Oxidative Stress and Apoptosis via Modulation of Mitochondrial Pathways in Broiler Chicken Kidneys. *Front. Vet. Sci.* **2021**, *8*, 750937. [CrossRef] [PubMed]
44. Li, L.Y.; Luo, X.; Wang, X. Endonuclease G is an apoptotic DNase when released from mitochondria. *Nature* **2001**, *412*, 95–99. [CrossRef] [PubMed]
45. Yuan, S.; Wu, B.; Yu, Z.; Fang, J.; Liang, N.; Zhou, M.; Huang, C.; Peng, X. The mitochondrial and endoplasmic reticulum pathways involved in the apoptosis of bursa of Fabricius cells in broilers exposed to dietary aflatoxin B1. *Oncotarget* **2016**, *7*, 65295–65306. [CrossRef]
46. Song, C.; Wang, Z.; Cao, J.; Dong, Y.; Chen, Y. Hesperetin protects hippocampal neurons from the neurotoxicity of Aflatoxin B1 in mice. *Ecotoxicol. Environ. Saf.* **2024**, *269*, 115782. [CrossRef] [PubMed]
47. Peng, X.; Yu, Z.; Liang, N.; Chi, X.; Li, X.; Jiang, M.; Fang, J.; Cui, H.; Lai, W.; Zhou, Y.; et al. The mitochondrial and death receptor pathways involved in the thymocytes apoptosis induced by aflatoxin B1. *Oncotarget* **2016**, *7*, 12222–12234. [CrossRef]
48. Lebeaupin, C.; Deborah, V.; Hazari, Y.; Hetz, C.; Bailly-Maitre, B. Endoplasmic Reticulum stress signaling and the pathogenesis of Non-Alcoholic Fatty Liver Disease. *J. Hepatol.* **2018**, *69*, 927–947. [CrossRef]
49. Wang, W.; Li, J.; Zhou, Q. The biological function of cytoplasm-translocated ENDOG (endonuclease G). *Autophagy* **2024**, *20*, 445–447. [CrossRef]
50. Tabas, I.; Ron, D. Integrating the mechanisms of apoptosis induced by endoplasmic reticulum stress. *Nat.Cell Biol.* **2011**, *13*, 184–190. [CrossRef]
51. Park, S.M.; Kang, T.I.; So, J.S. Roles of XBP1s in Transcriptional Regulation of Target Genes. *Biomedicines* **2021**, *9*, 791. [CrossRef]
52. Hillary, R.F.; FitzGerald, U. A lifetime of stress: ATF6 in development and homeostasis. *J. Biomed. Sci.* **2018**, *25*, 48. [CrossRef] [PubMed]
53. Iurlaro, R.; Muñoz-Pinedo, C. Cell death induced by endoplasmic reticulum stress. *FEBS J.* **2016**, *283*, 2640–2652. [CrossRef] [PubMed]
54. Tsukano, H.; Gotoh, T.; Endo, M.; Miyata, K.; Tazume, H.; Kadomatsu, T.; Yano, M.; Iwawaki, T.; Kohno, K.; Araki, K.; et al. he endoplasmic reticulum stress-C/EBP homologous protein pathway-mediated apoptosis in macrophages contributes to the instability of atherosclerotic plaques. *Arterioscler. Thromb. Vasc. Biol.* **2010**, *10*, 1925–1932. [CrossRef] [PubMed]
55. Liao, S.; Shi, D.; Clemons-Chevis, C.L.; Guo, S.; Su, R.; Qiang, P.; Tang, Z. Protective role of selenium on aflatoxin b1-induced hepatic dysfunction and apoptosis of liver in ducklings. *Biol. Trace. Elem. Res.* **2014**, *162*, 296–301. [CrossRef]
56. Murai, A.; Ebara, S.; Sasaki, S.; Ohashi, T.; Miyazaki, T.; Nomura, T.; Araki, S. Synergistic apoptotic effects in cancer cells by the combination of CLK and Bcl-2 family inhibitors. *PLoS ONE* **2020**, *15*, e0240718. [CrossRef] [PubMed]
57. Lakhani, S.A.; Masud, A.; Kuida, K.; Porter, G.A.J.; Booth, C.J.; Mehal, W.Z.; Inayat, I.; Flavell, R.A. Caspases 3 and 7: Key mediators of mitochondrial events of apoptosis. *Science* **2006**, *311*, 847–851. [CrossRef] [PubMed]
58. Yasin, M.; Mazdak, R.; Mino, I. Aflatoxin B1 impairs spermatogenesis: An experimental study for crosslink between oxidative stress and mitochondria-dependent apoptosis. *Environ. Toxicol.* **2018**, *33*, 1204–1213. [CrossRef]
59. Liu, Y.; Wang, W. Aflatoxin B1 impairs mitochondrial functions, activates ROS generation, induces apoptosis and involves Nrf2 signal pathway in primary broiler hepatocytes. *Anim. Sci. J.* **2016**, *87*, 1490–1500. [CrossRef] [PubMed]

veterinary sciences

MDPI

Case Report

Equine Herpesvirus-1 Induced Respiratory Disease in Dezhou Donkey Foals: Case Study from China, 2024

Lian Ruan [1,†], Liangliang Li [1], Rongze Yang [1,2,†], Anrong You [1,2], Muhammad Zahoor Khan [1], Yue Yu [1,3], Li Chen [1], Yubao Li [1], Guiqin Liu [1], Changfa Wang [1,*] and Tongtong Wang [1,*]

1 Liaocheng Research Institute of Donkey High-Efficiency Breeding and Ecological Feeding, Liaocheng University, Liaocheng 252000, China
2 College of Veterinary Medicine, Shanxi Agricultural University, Taigu 030801, China
3 College of Veterinary Medicine, Qingdao Agricultural University, Qingdao 266109, China
* Correspondence: wangchangfa@lcu.edu.cn (C.W.); wangtongtong@lcu.edu.cn (T.W.)
† These authors contributed equally to this work.

Simple Summary: Equine herpesvirus-1 (EHV-1) is known to cause severe respiratory diseases and abortions in horses. However, there is limited information regarding EHV-1 infection in donkeys, particularly in China. The present study reports the isolation of the virus from a one-month-old donkey foal (LC126) exhibiting severe respiratory disease. Genetic analysis revealed that the isolate is closely related to EHV-1. Notably, LC126 was also capable of inducing respiratory distress in a mouse model. These findings provide valuable insights into the pathogenicity of EHV-1 in donkeys.

Abstract: Equine herpesvirus-1 (EHV-1) is a significant pathogen that causes substantial economic losses in the equine industry worldwide, which leads to severe respiratory diseases and abortions in horses. However, reports of EHV-1 infection in donkeys are limited, particularly in China. This case study reported an EHV-1-induced respiratory disease in Dezhou donkey foals in Shandong Province, China, in July 2024. Three one-month-old foals exhibited high fever, nasal discharge, and respiratory distress, with a 100% mortality rate. The causative agent, strain LC126, was isolated from a one-month-old donkey foal exhibiting severe respiratory disease. Phylogenetic analysis of the EHV-1 isolate LC126 showed close similarity to EHV-1. Overall, our study revealed that EHV-1 can cause respiratory distress as well as death in donkeys. The study underscores the emerging threat of EHV-1 in donkeys and highlights the need for veterinarians and breeders to give proper attention to the potential threat of EHV-1 outbreaks.

Keywords: respiratory illness; donkey; EHV-1; phylogenetic analyses; pathogenicity

1. Introduction

Equid herpesviruses (EHVs) are a group of significant viral pathogens that pose a persistent threat to the equine industries worldwide [1,2]. To date, nine distinct equid herpesviruses (EHV-1–9) have been identified. Among these, EHV-1–5 primarily infects horses, while EHV-6–8, also known as asinine herpesvirus types 1–3 (AHV-1–3), predominantly affects donkeys [3,4]. EHV-1, first isolated in the United States in 1954, continues to be a widespread pathogen in equine populations globally [5]. EHV-1 is primarily associated with respiratory disorders, abortion, and viral encephalitis in horses [6].

In recent years, the growth of large-scale donkey farming in several regions in China has been accompanied by an increase in respiratory disease and abortion, both of which significantly impact the economic viability of the donkey industry. Interestingly, EHVs have been identified as key contributors to these health issues [7–9]. For instance, a study reported a strong correlation between EHV-1 and donkey abortions in Northern Xinjiang, China [10]. Furthermore, Wang et al. documented the association of EHV-8 with respiratory disease and miscarriages in donkeys [11,12]. Notably, reports of EHV-1-induced respiratory disease in donkeys have been rare. In this study, we present a comprehensive report of an EHV-1 infection in donkeys in Shandong Province, China, in 2024.

2. Materials and Methods

2.1. Herd History and Case Presentation

In July 2024, a respiratory disease outbreak occurred in a large-scale donkey farm located in Shandong Province, China. The affected animals were donkey foals, with three individuals presenting with severe clinical signs, including high fever (41 °C), nasal discharge, respiratory distress (dyspnea), depression, and anorexia. The duration of the illness was approximately five days, and the mortality rate was 100% among the affected foals. A one-month-old dead foal was submitted to the veterinary hospital of Liaocheng University for pathogen identification and diagnostic evaluation.

2.2. Nucleic Extraction, PCR Amplification and Sequencing

Nasal swabs and homogenized lung tissue samples from dead foals were collected and resuspended in 2 mL of phosphate-buffered saline (PBS). The samples underwent three freeze–thaw cycles. Subsequently, 1 mL of the PBS suspension was used for viral DNA extraction using the MiniBEST Viral RNA/DNA Extraction Kit (Takara Bio, Beijing, China). Polymerase Chain Reaction (PCR) was employed to detect EHV-1, EHV-4, and EHV-8 in the clinical samples. Specific primers (Table 1) were used for amplification, and PCR products were analyzed by 1% agarose gel electrophoresis. Additionally, another 1 mL of PBS was utilized to extract total RNA using SparkZol Reagent (Sparkjade, Jinan, China). The RNA was reversed into complementary DNA (cDNA) using Prime-Script™ Reverse Transcriptase (Takara Bio, Beijing, China). Furthermore, PCR was performed to detect influenza A virus H3N8 and equine arteritis virus (EAV) in the clinical samples with specific primers (Table 1), and the resulting PCR products were analyzed using the same method described above. Positive amplification products were subsequently sequenced by Sangon Biotech (Shanghai, China) for further analysis.

Table 1. The primer sequences in this study.

Primers	Primer Sequences (5′-3′)	PCR Product Sizes
EHV-1-*gB*-F	GAACCTCAGCCAACCCA	792 bp
EHV-1-*gB*-R	GCACTTTGCGGACGAAC	
EHV-4-*gB*-F	CTTAATCGCATTTAGACCGATG	1591 bp
EHV-4-*gB* R	CCGGAACTAGAAAGATGTTATGC	
EHV-8-*gG*-F	TCAGACTGTCACTCGTGGGA	316 bp
EHV-8-*gG*-R	CCTGGAGGCCGTTTAACACA	
EAV-*N*-F	ATGGCGTCAAGACGATCAC	333 bp
EAV-*N*-R	TTACGGCCCTGCTGGAGGC	
H3N8-*HA*-F	ATTCATCACCCGAGTTCAAA	595 bp
H3N8-*HA*-R	TTTCATTAATCTGGTCGATGGC	
H3N8-*M*-F	ACTGTGACAACCGAAGTG	224 bp
H3N8-*M*-R	TGCCTGGCCTTACTAGC	
EHV-1-*ORF33*-F	ATGTCCTCTGGTTGCCGTTC	2943 bp
EHV-1-*ORF33*-R	TTAAACCATTTTTCATTTTC	

2.3. Pathological Examination

Histopathological analysis was conducted as previously described [13]. In brief, lung tissues were collected from the dead donkey foals and fixed in 10% formalin solution. Following fixation, the tissues were dehydrated, embedded in paraffin, and sectioned into 4 mm thickness. The sections were then stained with hematoxylin and eosin (H&E) for general morphological evaluation. Immunohistochemistry (IHC) was performed to detect EHV-1 infection, utilizing a mouse anti-EHV-1 primary antibody, which was prepared in our laboratory. Finally, tissue samples were examined under a light microscope (Leica DMi8, Wetzlar, Germany), and representative images were captured for analysis.

2.4. Isolation of the Virus from Positive Lung Samples

The Rabbit kidney cells (RK-13) were seeded into 6-well plates and incubated overnight. Lung tissues were homogenized in a 5 mL centrifuge tube, followed by the addition of 3 mL of phosphate-buffered saline (PBS) containing penicillin and streptomycin. The tissue suspension was subjected to freeze–thaw cycles three times and subsequently filtered through a 0.22 μm sterilizing filter (Millipore, Merck, Darmstadt, Germany) after centrifugation. The resulting supernatant was then inoculated onto the RK-13 cell monolayer and incubated for 2 h. After the incubation period, the culture medium was replaced with minimum essential medium (MEM) supplemented with 3% fetal bovine serum (FBS) and incubated at 37 °C in a 5% CO_2 for three days. Cytopathic effects (CPEs) induced by the virus were monitored daily. The supernatants from the third passage were collected and analyzed for the presence of EHV-1 using PCR with specific primers (Table 1). The newly isolated EHV-1 strain was named LC126.

2.5. Morphological Observations Using Transmission Electron Microscope

The morphology of EHV-1 was observed as previously described [11]. In brief, the RK-13 cells were seeded into a T75 cm^2 cell culture flask overnight and infected with LC126 at 1 multiplicity of infection (MOI). These cells were fixed with TEM fixative at 48 hpi and collected by cell scraper and washed with PBS. This was followed by pre-embedding by 1% agarose and post-fixation in 1% osmium tetroxide in 0.1 M PB (pH 7.4) for 2 h at room temperature. Subsequently, dehydration in ethanol (30%, 50%, 70%, 80%, 95%, and 100%), the areas containing cells were block mounted, and the sections were cut into 60–80 nm sections using Leica ultramicrotome (Wetzlar, Germany). Finally, the sections were observed and imaged under a Hitachi transmission electron microscope (Tokyo, Japan).

2.6. Indirect Immunofluorescence

Immunofluorescence assays (IFA) were performed as described previously [14,15]. The coverslips were pre-seeded into 12-well cell plates overnight then infected with 0.1 MOI LC126. All cells were fixed at 48 hpi using 75% cold ethanol and sealed with 1% bull serum albumin (BSA), then these cells were incubated with primary antibody mouse EHV-1 convalescent serum and secondary antibodies with DyLight 488-conjugated affiniPure goat anti-mouse IgG + IgM (H+L) (Jackson, PA, USA). Finally, these cells were stained with 4,6-diamidino-2-phenylindole (DAPI) (Sparkjade, Jinan, China), and observed and captured by fluorescence microscope (Leica DMi8, Wetzlar, Germany).

2.7. The ORF33 of LC126 Amplification and Phylogenetic Analysis

The *ORF33* sequence from the LC126 isolate was amplified in this study; the reference sequences were downloaded from the GenBank database (http://www.ncbi.nlm.nih.gov/Genbank, accessed on January 2013). The phylogenetic tree was constructed using MEGA

6.0 software (Masatoshi Nei lab, State College, PA, USA) by the neighbor-joining method with 1000 bootstrap replications.

2.8. Infection of BALB/c

Ten female, specific pathogen-free BALB/c mice (6 weeks old) were obtained from Pengyue Laboratory Animal Breeding Co., Ltd. (Jinan, China) and randomly divided into infected group and mock group ($n = 5$). BALB/c were inoculated intranasally with LC126 (1×10^5 PFU/mice) or DMEM. Mice of each group were separated into different rooms to prevent cross-infection, and their body weights and clinical features were recorded. All mice were euthanized at 7 dpi via cervical dislocation, and the lung tissues were collected for pathologic analysis examination. Lung tissues of each group were fixed in 10% buffered formalin and embedded in paraffin. Paraffin-embedded tissue sections were used for H&E staining. In addition, the EHV-1 antigen in the lung tissue of different groups of mice was detected with anti-EHV-1 positive serum using IHC.

2.9. Statistical Analysis

Statistical analysis was performed using GraphPad Prism version 8.0 (San Diego, CA, USA). Differences between groups were compared using an unpaired Student's *t*-test. $p < 0.05$ was considered to indicate a statistically significant difference.

3. Results

3.1. Case Presentation and Pathogen Identification

In July 2024, three one-month-old donkey foals from a large-scale farm in Liaocheng, China, were observed to exhibit severe respiratory distress and subsequently died. The clinical signs, including anorexia, high fever, nasal discharge, dyspnea, and depression, were also observed in infected foals (Figure 1A). Necropsy of one deceased foal revealed a blue/purple discoloration of the lung tissue, with no significant gross abnormalities noted (Figure 1B). Histopathological examination of the lung showed hyperemia and hemorrhage of the alveolar septa (Figure 1C), as well as thickening of the alveolar septa and a notable increase in inflammatory cell infiltration (Figure 1D).

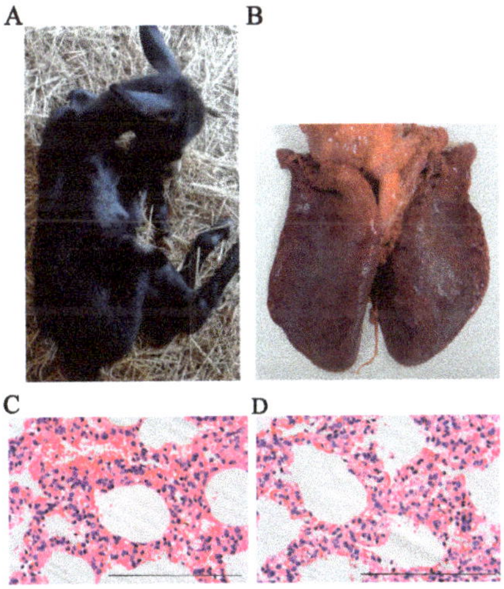

Figure 1. Gross lesions of a respiratory disease killed the foal: (**A**) A respiratory disease in foal donkey; (**B**) Gross change in lungs; (**C**) Hyperemia and hemorrhage in lung; (**D**) Severe interstitial pneumonia.

To identify the causative pathogen, lung tissue samples were processed and tested for various infectious agents. As shown in Figure 2, PCR testing for EAV (333 bp), H3N8 (595 bp or 224 bp), EHV-4 (1591 bp), and EHV-8 (316 bp) yielded negative results. However, the lung tissue tested positive for EHV-1, with a specific amplification of 792 bp corresponding to the EHV-1 *ORF70* gene.

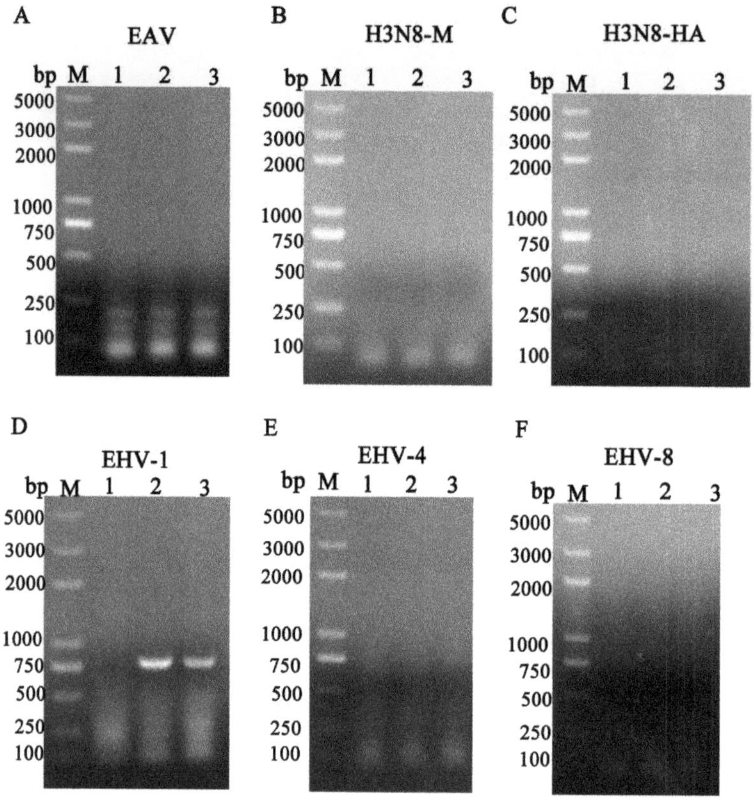

Figure 2. Screening of virus pathogens. Viral DNA/RNA was extracted from different samples: (**A**) EAV; (**B**) H3N8 M; (**C**) H3N8 HA; (**D**) EHV-1; (**E**) EHV-4; (**F**) EHV-8. They were detected by RT/PCR and PCR. Lane M represents a 5000 bp DNA molecular weight ladder. Moreover, 1 represents negative control, 2 represents nose swabs, and 3 represents lung of donkey foal.

Furthermore, the IHC staining was performed on lung tissue sections of donkey foal to further confirm EHV-1 infection. Positive staining for EHV-1 antigen was observed in the lung tissue, with brown-colored signals indicating the presence of the virus, as detected using mouse anti-EHV-1 serum (Figure 3A). No such positive signals were observed in the negative control section (Figure 3B).

3.2. *Identification of EHV-1 in the Isolates*

A field strain of EHV-1 was isolated from EHV-1-positive lung tissue on susceptible cell lines. Briefly, after a 3-day incubation period, CPEs were observed in RK-13 cells (Figure 4A, left) inoculated with the EHV-1-positive lung tissue, in contrast to mock-infected cells treated with PBS (Figure 4A, right). To confirm the presence of EHV-1 isolates, supernatants from CPE-positive cells were collected, and PCR assays were performed to detect EHV-1 across different passage cycles. As shown in Figure 4B, a single band of the expected size

(792 bp) was observed on a 1% agarose gel from passages P1 to P3 using specific primers for EHV-1 detection. Additionally, EHV-1 protein expression in RK-13 cells was confirmed by IFA assay using a mouse anti-EHV-1-positive serum. The isolates from the fourth passage exhibited virions both in the cytoplasm and nucleus (Figure 4C). Transmission electron microscopy (TEM) of EHV-1 virions revealed round particles approximately 110 nm in diameter, with a distinctive double-ring structure consisting of an inner nucleocapsid and an outer envelope (Figure 4D). This morphology is consistent with that of intact enveloped EHV-1 virions.

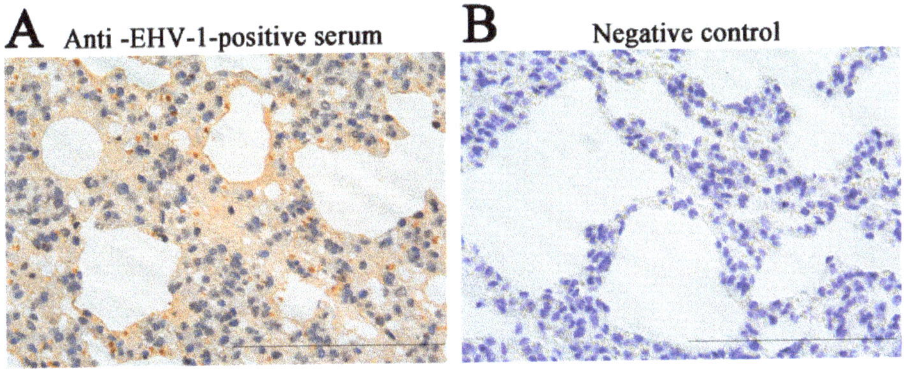

Figure 3. The immunohistochemistry (IHC) detection for EHV-1 in the lung of the donkey. The IHC was performed to detect the EHV-1 antigen in the lungs. The experimental group was treated with mouse anti-EHV-1 positive serum on the lung (**A**). The normal mouse serum-treated group served as a negative control on lung (**B**). Scale bars, 50 μm.

3.3. Phylogenetic Analysis of the ORF33 Gene

The complete *ORF33* gene sequence of LC126 was obtained through PCR amplification. Phylogenetic analysis revealed that this sequence shared the closest relationship with EHV-1, particularly the EHV-1 Ab1 and SDLC12 strains, and was distinct from other known EHV-4 and EHV-8 sequences (Figure 5).

3.4. Replication Characteristics and Pathogenicity of EHV-1 Isolate in BALB/C Mice

Previous studies have demonstrated that EHV-1 can effectively replicate in the lungs and brains of BALB/c mice [16,17]. In the present study, we also assessed the replication and virulence of the LC126 strain in BALB/c mice. As shown in Figure 6, mice infected with the LC126 strain exhibited clinical signs, including depression, ruffled fur, inappetence, respiratory distress, and crouching in corners by 3 dpi, in contrast to the mock-infected group (Figure 6A). Additionally, a significant decrease in body weight was observed in the LC126-infected mice starting at 3 dpi (Figure 6B). Viral titers in the lungs of infected mice were quantified, with a mean titer of $10^{3.9}$TCID$_{50}$ (Figure 6C). Histopathological analysis revealed severe lesions in the lung tissue, characterized by hyperemia and hemorrhage in LC126-infected mice (Figure 6D, left). The IHC staining was performed to detect EHV-1 antigen in the lung tissue of LC126-infected mice, using anti-EHV-1 positive serum, which confirmed the presence of viral antigen (Figure 6D, right).

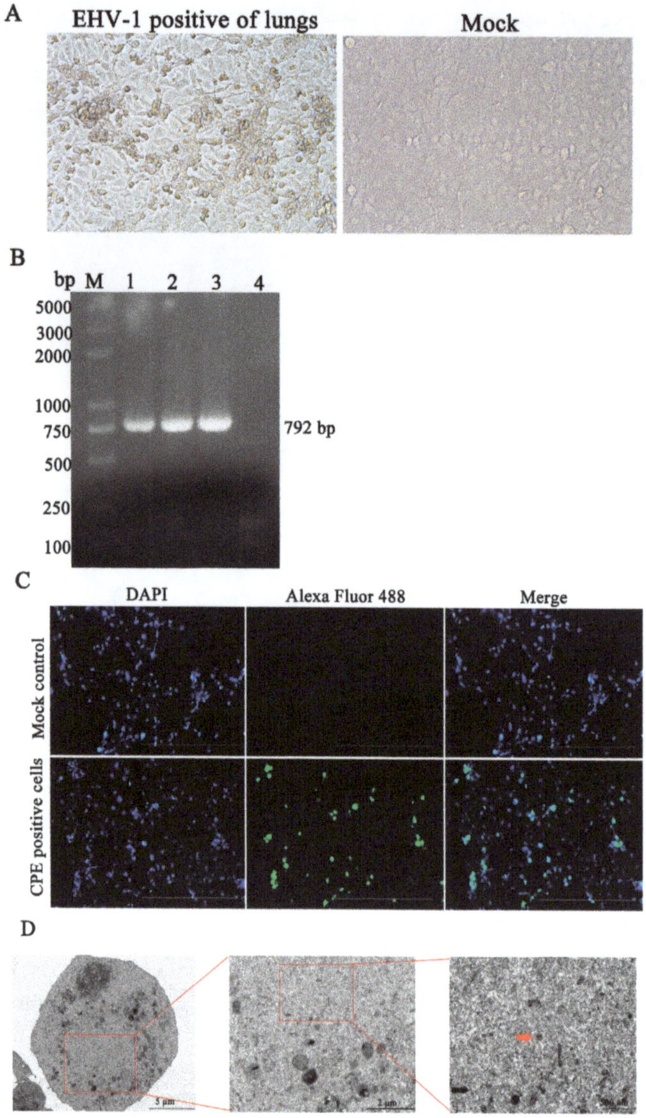

Figure 4. Identification of EHV-1 isolation. The RK-13 cells were inoculated with the supernatant of EHV-1-positive lung tissue (left panel) or mock control (right panel) (**A**). The cytopathogenic effect (CPE) was observed using microscopy at 48 h pi. Scale bars, 100 mm. (**B**) The gB gene of the EHV-1 isolate was confirmed by PCR. The PCR products were analyzed by 1% agarose gel. Regarding the DL2000 plus DNA marker (lane M), 1 represents EHV-1 Passage 1 cell supernatant, 2 represents EHV-1 Passage 2 cell supernatant, 3 represents EHV-1 passage 3 cell supernatant, and 4 represents negative control. (**C**) The isolate was detected by IFA. The images represent the subcellular locations of EHV-1 proteins using IFA detection with anti-EHV-1 mouse serum and the corresponding alexa fluor 488-conjugated secondary antibodies. Cells were imaged by Leica DMi8. Scale bars, 50 μm. (**D**) Transmission electron micrograph analysis. RK-13 cells were infected with LC126 (MOI = 1) and then fixed by TEM fixative at 48 hpi and observed by transmission electron microscopy. Magnified images of the regions indicated by red rectangles are demonstrated on the right. Red arrows represent EHV-1 virions.

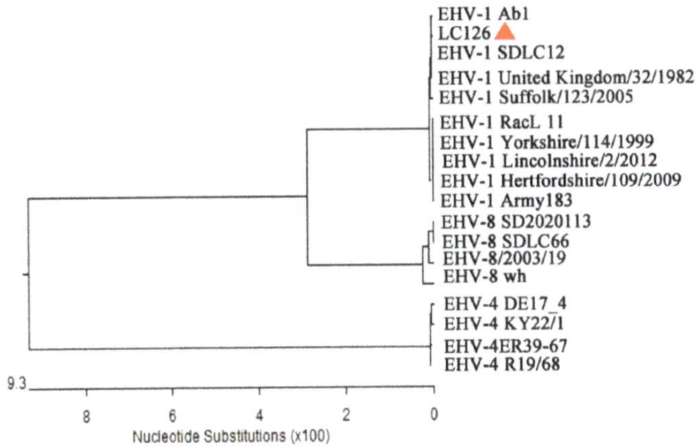

Figure 5. Phylogenetic tree based on the sequence of *ORF33* from the isolate in the present study and with known sequences of EHVs. Sequences of isolates of *ORF33* in the present study are labeled with "▲". The scale bar indicates nucleotide substitutions per site.

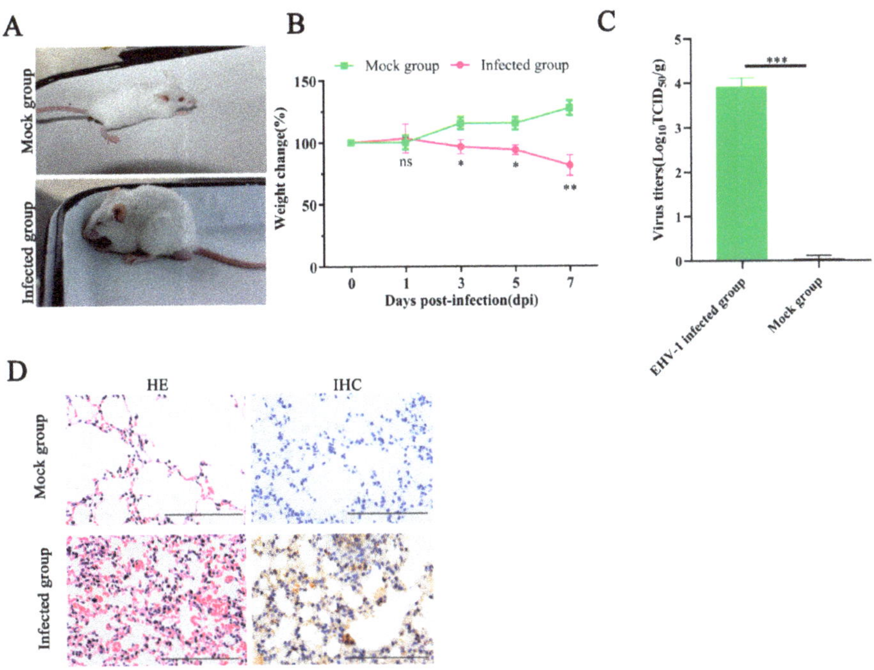

Figure 6. The pathogenicity of LC126 in Balb/c mice. Ten female, specific pathogen-free BALB/c mice (6 weeks old) were randomly divided into an infected group and a mock group (n = 5). Clinical signs (**A**), the percent change in body weight was calculated for each mouse based on the initial starting weight before virus inoculation (**B**). The lung tissues of the infected group and mock group were collected at 7 dpi to evaluate virus titers using $TCID_{50}$. * $p < 0.05$; ** $p < 0.01$; *** $p < 0.001$. (**C**). The lung tissues of different groups of mice were collected at 7 dpi and fixed in 10% formalin solution for H&E detection (**D**, left) and IHC analysis (**D**, right).

4. Discussion

Large-scale donkey farming has seen significant growth in China in recent years, contributing to the development of the animal husbandry sector [18,19]. However, the sustainable development of the donkey industry is under threat due to the prevalence of respiratory diseases and abortion, which present considerable challenges to its continued expansion [20,21]. Notably, respiratory diseases in donkeys have been linked to various microbial agents. For instance, a study by Yang et al. [22] identified the equine influenza virus H3N8 as a significant cause of respiratory illness in donkeys. In addition, a study conducted in Shandong Province, China, reported a 21.5% positive rate (14/65) of equine influenza virus H3N8 among donkeys exhibiting respiratory diseases on large-scale farms [23]. A major concern in equine respiratory health is the role of EHV, particularly EHV-1 and EHV-4, which are well-established pathogens associated with respiratory tract infections in horses [24,25]. In addition to these viruses, several studies have implicated EHV-8 in respiratory diseases and abortion in both horses and donkeys [4,11,26]. Moreover, EAV has been recognized as a causative agent of both respiratory and reproductive diseases in equids [27–29]. The above-mentioned viruses were identified in the dead donkey foals. In this instance, H3N8, EAV, EHV-4, and EHV-8 were negative; only EHV-1 appears positive (Figure 2). Moreover, respiratory diseases in the donkey industry have been linked to bacterial pathogens, like *Streptococcus subsp. Equi* and *Rhodococcus equi*. In the present study, we first performed detection of these bacteria, but they are negative. Only EHV-1 was identified as the sole positive pathogen among the potential viral agents. While H3N8, EAV, EHV-4, and EHV-8 were all tested negative, EHV-1 was found to be present in the lung tissue of dead donkey foals (Figures 2 and 3). This finding suggests that EHV-1 may be a primary pathogenic agent contributing to respiratory disease in Dezhou donkey foals (Figure 4).

Previous studies showed that EHV-1 was identified from donkeys, both with and without respiratory illness, in Ethiopia [30,31]. More recently, EHV-1 was isolated from the lung tissue of an aborted fetal donkey, further supporting its potential role in reproductive and respiratory diseases in donkeys [10]. Our study represents the first report of EHV-1 as the causative agent of respiratory disease and death in Dezhou donkey foals in Shandong Province, China. Histopathological examination of affected donkey foals revealed that EHV-1 infection led to severe alveolar swelling, inflammatory cell infiltration, and hemorrhage in the lungs (Figures 1 and 3). It is noteworthy that the mice model was widely used as an animal model to detect EHVs in pathogenicity studies [3,17,32,33]. In the present study, similar pathological changes were validated and observed in a mouse model (Figure 6). Collectively, these results highlight the pathogenicity of EHV-1 in donkeys and suggest that this virus contributes significantly to respiratory disease and mortality in donkey foals. Given the impact of EHV-1 on donkey health, particularly in large-scale farming operations, it is imperative to invest in further research into effective preventative measures.

5. Conclusions

This study presented a case report on the EHV-1-induced infection in large-scale donkey farms in China. Our findings demonstrate that EHV-1 infection is clinically associated with respiratory distress and mortality in donkey foals. EHV-1 also causes abortion in pregnant donkeys. Given the significant impact of EHV-1 on donkey health, it is crucial for the donkey industry to implement robust biosafety management strategies to prevent the spread of this virus. Enhanced surveillance and improved biosecurity protocols are essential for mitigating the risks posed by EHV-1 and ensuring the sustainability of the donkey farming industry.

Author Contributions: L.R., L.L. and R.Y.: conceptualization, data curation, formal analysis, investigation, methodology, validation, visualization, writing—original draft, writing—review and editing. A.Y., M.Z.K., Y.Y. and L.C.: data curation, formal analysis, investigation, methodology and writing—review and editing. Y.L. and G.L.: writing—review and editing, validation, resources, visualization.

C.W. and T.W.: conceptualization, funding acquisition, validation, methodology, project administration, resources, software, supervision, writing—original draft, writing—review and editing. All authors have read and agreed to the published version of the manuscript.

Funding: This work was supported by grants from Natural Science Foundation of Shandong Province (ZR2024MC162), the Project of Shandong Province Higher Educational Science and Technology Program for Youth (2022KJ287), Shandong Province Agricultural Major Technology Collaborative Promotion Plan (SDNYXTTG-2024-13), the Shandong Province Modern Agricultural Technology System Donkey Industrial Innovation Team (SDAIT-27, SDAIT-27-18), and the Open Project of Liaocheng University Animal Husbandry Discipline (319312105-25, and 319312105-26).

Institutional Review Board Statement: All procedures were approved by the Animal Welfare and Ethics Committee of the Institute of Animal Science, Liaocheng University (protocol number LC2024-07) and performed according to the Animal Ethics Procedures and Guidelines of the Ministry of Agriculture of China. Nasal swabs and lungs were collected, and the animal experiments were conducted, according to the approved procedures. Written consent regarding the donkeys was obtained from the farm owners.

Informed Consent Statement: Informed consent was obtained from all subjects involved in the study.

Data Availability Statement: The data supporting this study's findings are available from the corresponding author upon reasonable request.

Conflicts of Interest: The authors declare that they have no competing interests.

References

1. Afify, A.F.; Hassanien, R.T.; El Naggar, R.F.; Rohaim, M.A.; Munir, M. Unmasking the ongoing challenge of equid herpesvirus-1 (EHV-1): A comprehensive review. *Microb. Pathog.* **2024**, *193*, 106755. [CrossRef]
2. Tau, R.L.; Ferreccio, C.; Bachir, N.; Torales, F.; Romera, S.A.; Maidana, S.S. Comprehensive analysis of equid herpesvirus recombination: An insight into the repeat regions. *J. Equine Vet. Sci.* **2023**, *130*, 104916. [CrossRef] [PubMed]
3. Wang, T.; Hu, L.; Liu, M.; Wang, T.; Hu, X.; Li, Y.; Liu, W.; Li, Y.; Wang, Y.; Ren, H.; et al. The emergence of viral encephalitis in donkeys by equid herpesvirus 8 in China. *Front. Microbiol.* **2022**, *13*, 840754. [CrossRef]
4. Garvey, M.; Suarez, N.M.; Kerr, K.; Hector, R.; Moloney-Quinn, L.; Arkins, S.; Davison, A.J.; Cullinane, A. Equid herpesvirus 8: Complete genome sequence and association with abortion in mares. *PLoS ONE* **2018**, *13*, e0192301. [CrossRef]
5. Doll, E.R.; Wallace, E.; Richards, M.G. Thermal, hematological, and serological responses of weanling horses following inoculation with equine abortion virus: Its similarity to equine influenza. *Cornell Vet.* **1954**, *44*, 181–190. [PubMed]
6. Oladunni, F.S.; Horohov, D.W.; Chambers, T.M. EHV-1: A constant threat to the horse industry. *Front. Microbiol.* **2019**, *10*, 2668. [CrossRef] [PubMed]
7. Chen, L.; Li, S.; Li, W.; Yu, Y.; Sun, Q.; Chen, W.; Zhou, H.; Wang, C.; Li, L.; Xu, M.; et al. Rutin prevents EqHV-8-induced infection and oxidative stress via Nrf2/HO-1 signaling pathway. *Front. Cell. Infect. Microbiol.* **2024**, *14*, 1386462. [CrossRef]
8. Li, L.; Cui, X.; Yu, Y.; Sun, Q.; Li, W.; Li, Y.; Li, S.; Chen, L.; Khan, M.Z.; Wang, C.; et al. Blebbistatin as a novel antiviral agent targeting equid herpesvirus type 8. Front. *Vet. Sci.* **2024**, *11*, 1390304.
9. Li, S.; Li, L.; Sun, Y.; Khan, M.Z.; Yu, Y.; Ruan, L.; Chen, L.; Zhao, J.; Jia, J.; Li, Y.; et al. Protective role of cepharanthine against equid herpesvirus type 8 through AMPK and Nrf2/HO-1 pathway activation. *Viruses* **2024**, *16*, 1765. [CrossRef]
10. Tong, P.; Pan, J.; Dang, Y.; Yang, E.; Jia, C.; Duan, R.; Tian, S.; Palidan, N.; Kuang, L.; Wang, C.; et al. First identification and isolation of equine herpesvirus type 1 in aborted fetal lung tissues of donkeys. *Virol. J.* **2024**, *21*, 117. [CrossRef]
11. Wang, T.; Xi, C.; Yu, Y.; Liu, W.; Akhtar, M.F.; Li, Y.; Wang, C.; Li, L. Characteristics and epidemiological investigation of equid herpesvirus 8 in donkeys in Shandong, China. *Arch. Virol.* **2023**, *168*, 99. [CrossRef]
12. Wang, T.; Hu, L.; Wang, Y.; Liu, W.; Liu, G.; Zhu, M.; Zhang, W.; Wang, C.; Ren, H.; Li, L. Identification of equine herpesvirus 8 in donkey abortion: A case report. *Virol. J.* **2022**, *19*, 10. [CrossRef]
13. Wang, T.; Du, Q.; Niu, Y.; Zhang, X.; Wang, Z.; Wu, X.; Yang, X.; Zhao, X.; Liu, S.L.; Tong, D.; et al. Cellular p32 is a critical regulator of porcine circovirus type 2 nuclear egress. *J. Virol.* **2019**, *93*, e00979-19. [CrossRef] [PubMed]
14. Li, L.; Wang, J.; Chen, L.; Ren, Q.; Akhtar, M.F.; Liu, W.; Wang, C.; Cao, S.; Liu, W.; Zhao, Q.; et al. Diltiazem HCl suppresses porcine reproductive and respiratory syndrome virus infection in susceptible cells and in swine. *Vet. Microbiol.* **2024**, *292*, 110054. [CrossRef]

15. Wang, T.; Hu, L.; Li, R.; Ren, H.; Li, S.; Sun, Q.; Ding, X.; Li, Y.; Wang, C.; Li, L. Hyperoside inhibits EHV-8 infection via alleviating oxidative stress and IFN production through activating JNK/Keap1/Nrf2/HO-1 signaling pathways. *J. Virol.* **2024**, *98*, e0015924. [CrossRef] [PubMed]
16. Fuentealba, N.A.; Sguazza, G.H.; Zanuzzi, C.N.; Bravi, M.E.; Scrochi, M.R.; Valera, A.R.; Corva, S.G.; Gimeno, E.J.; Pecoraro, M.R.; Galosi, C.M. Immunoprotective response induced by recombinant glycoprotein D in the BALB/c respiratory mouse model of Equid alphaherpesvirus 1 infection. *Rev. Argent. Microbiol.* **2019**, *51*, 119–129. [CrossRef]
17. Walker, C.; Ruitenberg, K.M.; Love, D.N.; Millar Whalley, J. Immunization of BALB/c mice with DNA encoding equine herpesvirus 1 (EHV-1) glycoprotein D affords partial protection in a model of EHV-1-induced abortion. *Vet. Microbiol.* **2000**, *76*, 211–220. [CrossRef]
18. Zhang, Z.; Huang, B.; Wang, Y.; Zhu, M.; Liu, G.; Wang, C. A survey report on the donkey original breeding farms in China: Current aspects and future prospective. *Front. Vet. Sci.* **2023**, *10*, 1126138. [CrossRef] [PubMed]
19. Seyiti, S.; Kelimu, A. Donkey industry in China: Current aspects, suggestions and future challenges. *J. Equine Vet. Sci.* **2021**, *102*, 103642. [CrossRef]
20. Li, L.; Li, S.; Ma, H.; Akhtar, M.F.; Tan, Y.; Wang, T.; Liu, W.; Khan, A.; Khan, M.Z.; Wang, C. An overview of infectious and non-infectious causes of pregnancy losses in equine. *Animals* **2024**, *14*, 1961. [CrossRef]
21. Rickards, K.J.; Thiemann, A.K. Respiratory disorders of the donkey. *Vet. Clin. North Am. Equine Pract.* **2019**, *35*, 561–573. [CrossRef]
22. Yang, H.; Xiao, Y.; Meng, F.; Sun, F.; Chen, M.; Cheng, Z.; Chen, Y.; Liu, S.; Chen, H. Emergence of H3N8 equine influenza virus in donkeys in China in 2017. *Vet. Microbiol.* **2018**, *214*, 1–6. [CrossRef] [PubMed]
23. Yongfeng, Y.; Xiaobo, S.; Nan, X.; Jingwen, Z.; Wenqiang, L. Detection of the epidemic of the H3N8 subtype of the equine influenza virus in large-scale donkey farms. *Int. J. Vet. Sci. Med.* **2020**, *8*, 26–30. [CrossRef]
24. Pusterla, N.; Sandler-Burtness, E.; Barnum, S.; Hill, L.A.; Mendonsa, E.; Khan, R.; Portener, D.; Ridland, H.; Schumacher, S. Frequency of detection of respiratory pathogens in nasal secretions from healthy sport horses attending a spring show in California. *J. Equine Vet. Sci.* **2022**, *117*, 104089. [CrossRef] [PubMed]
25. Pusterla, N.; Leutenegger, C.M.; Barnum, S.; Wademan, C.; Hodzic, E. Challenges in navigating molecular diagnostics for common equine respiratory viruses. *Vet. J.* **2021**, *276*, 105746. [CrossRef]
26. Liu, C.; Guo, W.; Lu, G.; Xiang, W.; Wang, X. Complete genomic sequence of an equine herpesvirus type 8 Wh strain isolated from China. *J. Virol.* **2012**, *86*, 5407. [CrossRef]
27. Otzdorff, C.; Beckmann, J.; Goehring, L.S. Equine arteritis virus (EAV) outbreak in a show stallion population. *Viruses* **2021**, *13*, 2142. [CrossRef]
28. Rivas, J.; Neira, V.; Mena, J.; Brito, B.; Garcia, A.; Gutierrez, C.; Sandoval, D.; Ortega, R. Identification of a divergent genotype of equine arteritis virus from South American donkeys. *Transbound. Emerg. Dis.* **2017**, *64*, 1655–1660. [CrossRef]
29. Stadejek, T.; Mittelholzer, C.; Oleksiewicz, M.B.; Paweska, J.; Belak, S. Highly diverse type of equine arteritis virus (EAV) from the semen of a South African donkey: Short communication. *Acta Vet. Hung.* **2006**, *54*, 263–270. [CrossRef]
30. Negussie, H.; Gizaw, D.; Tesfaw, L.; Li, Y.; Oguma, K.; Sentsui, H.; Tessema, T.S.; Nauwynck, H.J. Detection of equine herpesvirus (EHV)-1, -2, -4, and -5 in Ethiopian equids with and without respiratory problems and genetic characterization of EHV-2 and EHV-5 strains. *Transbound. Emerg. Dis.* **2017**, *64*, 1970–1978. [CrossRef]
31. Temesgen, T.; Getachew, Y.; Negussie, H. Molecular identification of equine herpesvirus 1, 2, and 5 in equids with signs of respiratory disease in central Ethiopia. *Vet. Med.* **2021**, *12*, 337–345. [CrossRef]
32. Mesquita, L.P.; Arévalo, A.F.; Zanatto, D.A.; Miyashiro, S.I.; Cunha, E.M.S.; de Souza, M.D.C.C.; Villalobos, E.M.C.; Mori, C.M.C.; Maiorka, P.C.; Mori, E. Equine herpesvirus type 1 induces both neurological and respiratory disease in Syrian hamsters. *Vet. Microbiol.* **2017**, *203*, 117–124. [CrossRef] [PubMed]
33. Abas, O.; Abdo, W.; Kasem, S.; Alwazzan, A.; Saleh, A.G.; Saleh, I.G.; Fukushi, H.; Yanai, T.; Haridy, M. Time Course-Dependent Study on Equine Herpes Virus 9-Induced Abortion in Syrian Hamsters. *Animals* **2020**, *10*, 1369. [CrossRef] [PubMed]

veterinary sciences

MDPI

Article

Evaluating Forelimb and Hindlimb Joint Conformation of Morna Racehorses (*Equus caballus*)

Israr Ahmad [1], Sahar Ijaz [2,*], Mirza M. Usman [2], Ayesha Safdar [3], Imdad U. Khan [4], Muhammad Zeeshan [5] and Syed S. U. H. Bukhari [5,*]

1 Department of Basic Veterinary Sciences, Faculty of Veterinary and Animal Sciences, Gomal University, Dera Ismail Khan 29220, Pakistan; israrahmad@gu.edu.pk
2 Department of Anatomy and Histology, University of Veterinary and Animal Sciences, Lahore 54000, Pakistan; mirza.usman@uvas.edu.pk
3 Department of Surgery, University of Veterinary and Animal Sciences, Lahore 54000, Pakistan; ayesha.safdar@uvas.edu.pk
4 Department of Clinical Sciences, Faculty of Veterinary and Animal Sciences, Gomal University, Dera Ismail Khan 29220, Pakistan; imdadsaifi@gmail.com
5 Jockey Club College of Veterinary Medicine and Life Sciences, City University of Hong Kong, Hong Kong SAR 523808, China; mzeeshan6-c@my.cityu.edu.hk
* Correspondence: sahar.ijaz@uvas.edu.pk (S.I.); habukhari2-c@my.cityu.edu.hk (S.S.U.H.B.)

Simple Summary: The assessment of limb joint angles is critical for understanding horse conformation, performance, and safety. This study aimed to determine the normal limb joint angles of the Morna breed, a Pakistani racehorse that has received little academic attention. We used standard techniques to measure joint angles of the forelimbs and hindlimbs. Morna racehorses had elbow and stifle joint angles correlated with peak racing performance, but their hock and fetlock angles differed from other breeds, for example, jumping Thoroughbreds and French trotters. This underscores the necessity for additional research into the Morna breed's conformation, such as measuring limb segment lengths and comparing them to joint angles. Overall, quantifying joint angles in horses, especially in understudied breeds such as the Morna racehorses, is critical for understanding conformation, performance, and injury prevention.

Abstract: Measuring limb joint angles is crucial for understanding horse conformation, performance, injury diagnosis, and prevention. While Thoroughbred horses have been extensively studied, local Pakistani breeds (e.g., Morna racehorse) have not received scientific attention. This study aimed to quantify normal angles of limb joints in the Morna breed. Limb joint angles of standing horses (n = 50) were quantified using a measuring tape, height stick, protractor scale, and goniometer. The mean and standard deviation (Mean $\pm$ SD) values for the forelimb joint angles were $123.02 \pm 3.46^\circ$ for elbow, $171.52 \pm 2.39^\circ$ for knee, and $147.68 \pm 5.11^\circ$ for fetlock. The mean $\pm$ SD values for the hindlimb joint angles were $128.62 \pm 4.08^\circ$ for stifle, $160.40 \pm 3.89^\circ$ for hock, and $155.48 \pm 2.68^\circ$ for fetlock. There was a non-significant ($p > 0.05$) correlation between horse joint angles and, age, body weight, and body condition score (BCS). The elbow and stifle joint angles of Morna align well with characteristics associated with optimal racing performance. However, the hock and fetlock angles differ from jumping Thoroughbred and French trotters. We recommend further research to examine the conformation of the Morna breed, particularly by measuring the lengths of limb segments and correlating it with joint angles. This may provide valuable insights into individual variations within the breed.

Keywords: body condition score; equine biomechanics; goniometer; horse: joint anatomy; joint conformation; limb joint angles; measuring joint angles; Morna breed; racehorses

1. Introduction

Horseracing is an ancient sport that has evolved alongside human civilization [1]. Successful horseracing requires horses with high-speed running abilities, largely determined by each horse's conformation and limb soundness [2]. It is common for racehorses to develop soundness issues at a young age, often due to poor skeletal conformation [3]. In racehorses, the most likely anatomical locations for injuries can often be predicted based on the horse's conformation [2]. Several factors influence a horse's conformation, including genetics, racing surface, prior injuries or pathologies, biomechanics, and the horse's age [3]. Horse stride problems occur due to anatomically compromised limbs, such as gait asymmetry, and unstable gait [4,5]. To assess whether an athletic horse is likely to succeed or fail, it is essential to evaluate its conformation in relation to speed [6]. Biomechanical limitations are associated with equine musculoskeletal disorders, such as angular limb deformities. Key conformational traits of the limbs include leg stance, hoof quality, and joint angles [6].

Limbs play a crucial role in movement efficiency and performance in race events. The structure and alignment of the leg bones directly influence the horse's function [6]. Racehorses, in particular, place extra strain on their hindlimbs, relying on their hindquarters for engagement and collection [7,8]. Amid higher activity in the hindlimbs and the pelvic area, the soft tissues, ligaments, and joints are at a higher risk for injuries [7–9]. Additionally, the forelimbs experience substantial load during take-off and landing, carrying the horse's total weight upon landing [10]. Consequently, the distal limb joints and surrounding soft tissues face extra stress and weight [11]. The rate of progression is directly linked to the health of the musculoskeletal system [2]. Poor conformation often indicates defects in the tendons and ligaments, leading to improper joint angles [12].

Measuring the angles of limb joints is a crucial aspect of studying horse conformation, as it plays a significant role in evaluating performance and selecting a high-performance racehorse [13,14]. It also helps prevent injuries, diagnose lameness, monitor training and rehabilitation, and make informed breeding and horse selection decisions [2,15–17]. Understanding limb joint angles provides valuable insights into a horse's musculoskeletal health, enabling proactive management strategies that ensure optimal performance, reduce the risk of injuries, and promote long-term health and welfare [14,18–20].

Conformation evaluation is subjective and relies on the assessor's experience, which can vary among experts [14,20,21]. For the most accurate assessment, a horse's limb conformation should be evaluated while it stands straight on a hard, level surface under the observation of an experienced coach [16]. Every region has its evaluation procedures based on specific standards and breed-specific laws. While extensive research was conducted on angulation and conformation in long-distance racehorse breeds [14,16,20], local and short-distance Pakistani racehorses have received little attention.

This study aimed to evaluate and report normal angles of limb joints in Pakistan's Morna breed of racehorses and establish baseline measurements for joint conformation. We hypothesized that documenting normal angles in healthy horses will create standard benchmarks for Morna racehorses and assist local researchers in conducting further biomechanical studies on this lesser-known horse breed.

2. Materials and Methods

2.1. Study Area and Study Design

This study was conducted in southern Khyber Pakhtunkhwa, Pakistan, involving active Morna breed racehorses (four to six years of age) with a good body condition score (four to six out of nine). Data were collected from 50 male racehorses. Before the start of data collection, written consent was obtained from the horse owners. Horses with a history of bone fractures, chronic musculoskeletal diseases, or any musculoskeletal defects and medication were excluded from this study.

2.2. Measurements

Measurements were taken on racehorses while they stood in a normal position (Figure 1). We used a measuring tape, height stick, goniometer, and protractor scale and established predetermined reference points for accuracy [16], (Figure 1). Goniometry was used for angle measurement, as it is simple, practical, cost-effective, less labor-intensive, repeatable, and reliable [22]. This method does not require sophisticated equipment and is widely used in conformational studies [22–24]. This is particularly significant in low-resource settings where access to advanced technology may be limited. Before determining the joint angles, we established reference points based on their full range of motion to identify the axis of rotation for each joint. We selected easily palpable landmarks that had been used in previous conformational studies [16], (Figure 1). Data collection was conducted on the left side of the animals, and all the measurements were taken by a single researcher.

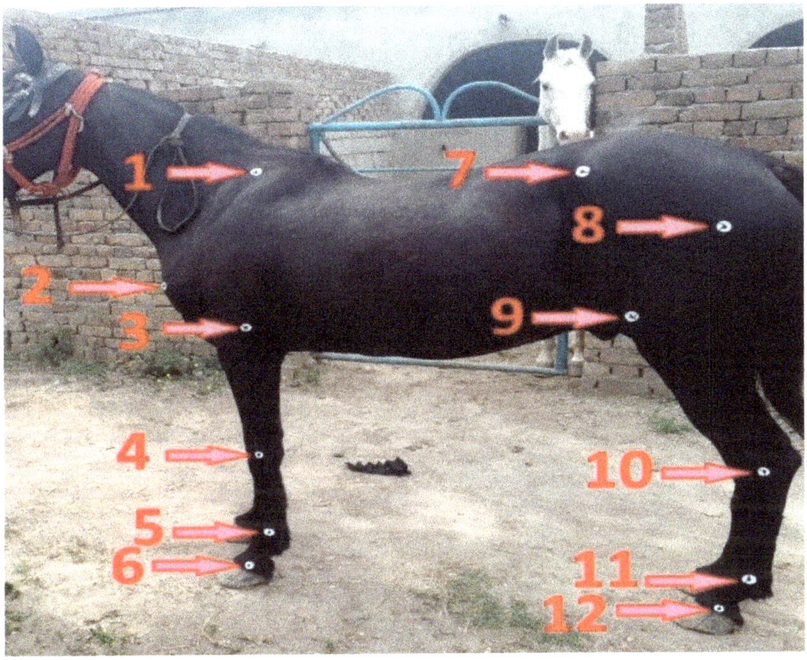

Figure 1. Standard joint positions and body reference points. The numbered reference points mentioned in the figure show the middle of the (1) dorsal border of the scapula, (2) point of the shoulder, (3) elbow joint, (4) knee joint, (5) fore fetlock joint, (6) fore pastern joint, (7) tuber coxae, (8) hip joint, (9) stifle joint, (10) hock joint, (11) hind fetlock joint, and (12) hind pastern joint.

We quantified shoulder versus ground angle, shoulder joint angle, elbow joint angle, knee joint angle, fore fetlock joint angle, fore pastern versus ground angle, hip joint angle, stifle joint angle, hock joint angle, hind fetlock joint angle, and hind pastern versus ground angle according to existing studies [14,16]. The data on all angles was collected using a measuring tape, a height stick, a protractor scale, and a goniometer. The length of the goniometer's arms was modified according to the researcher's needs; for example, for carpal and tarsal measurements, the arms of the goniometer were extended, and for metacarpophalangeal and metatarsophalangeal measurements, the arms were shortened.

Body condition scores (BCS) of all horses were measured, and for weight determination, girth circumference was measured by measuring tape, and a height stick measured body length. The weight of the horse was measured with the help of the following formula [25]:

$$\text{Horse body weight} = ((\text{Girth}^2\ (\text{cm}) \times \text{Body length (cm)})/2.54^3/660 + 22.7$$

2.3. *Data Collection*

2.3.1. Shoulder Versus Ground Angle

A straight line was drawn from the middle of the dorsal border of the scapula down to the point of the shoulder using a height stick. From the point of the shoulder, the line was extended distally and cranially toward the ground until it contacted the ground surface. Another straight line was drawn horizontally on the ground. The angle was measured by placing the protractor scale directly on the ground at the point where the lines from the shoulder and the ground intersect (Figure 2).

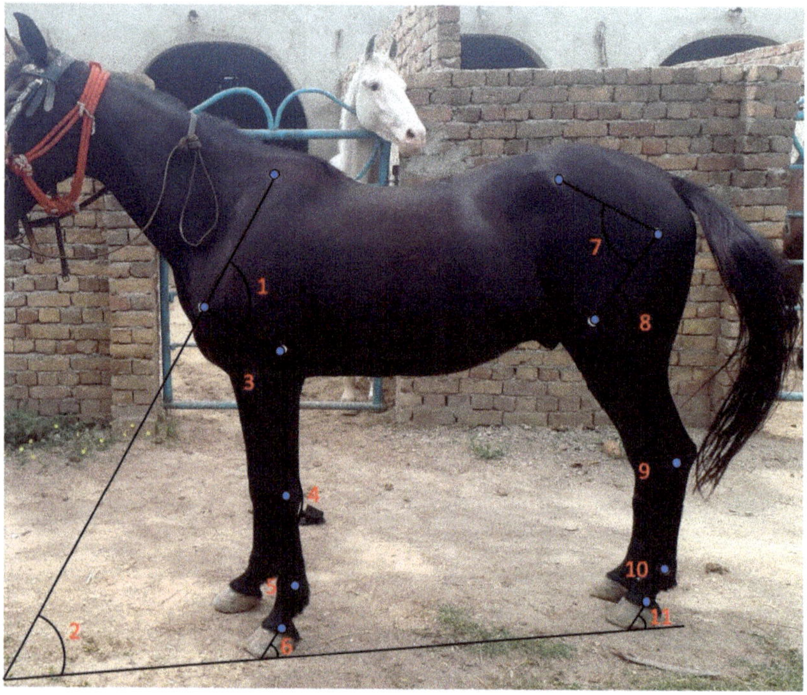

Figure 2. Studied angles: (1) shoulder joint angle; (2) shoulder versus ground angle; (3) elbow joint angle; (4) knee joint angle; (5) fore fetlock joint angle; (6) fore pastern versus ground angle; (7) hip joint angle; (8) stifle joint angle; (9) hock joint angle; (10) hind fetlock joint angle; and (11) hind pastern versus ground angle.

2.3.2. Shoulder Joint Angle

A straight line was drawn from the middle of the dorsal border of the scapula down to the point of the shoulder using a height stick. Another straight line was drawn from the middle of the shoulder joint toward the middle of the elbow joint. A protractor was positioned at the shoulder joint, ensuring it was angled in relation to the line leading from the shoulder to the elbow, and the angle was measured with a goniometer. The disc of the goniometer was placed at the middle of the shoulder joint, with its fixed arm aligned along the spine of the scapula, while the movable arm extended toward the midline of the humerus, pointing toward the middle of the elbow joint (Figure 2).

2.3.3. Elbow Joint Angle

A straight line was drawn from the middle of the shoulder joint to the middle of the elbow joint. Another straight line was drawn from the elbow joint to the middle of the knee joint using a height stick. A protractor was placed at the middle of the elbow joint, with one line extending toward the shoulder and the other toward the knee. The angle was measured using a goniometer, and the disc was placed at the axis of the elbow joint. One arm of the goniometer was aligned with the line from the shoulder to the elbow joint, while the other arm extended from the elbow to the knee joint (Figure 2).

2.3.4. Knee Joint Angle

A straight line was drawn from the middle of the elbow joint to the middle of the knee joint. Another line extended from the middle of the knee joint to the middle of the fore fetlock joint. A protractor was placed obliquely at the knee joint. The angle was measured using a goniometer, with one arm positioned at the middle of the line extending from the elbow joint proximally and the other arm aligned with the line extending from the knee to the fore fetlock distally (Figure 2).

2.3.5. Fore Fetlock Joint Angle

A straight line was drawn from the knee joint's midpoint to the fore fetlock joint. Another line was drawn from the middle of the fore fetlock joint to the middle of the fore pastern joint. A protractor was placed over the fore fetlock joint, aligning it with the junction of the proximal and distal lines. The angle was measured using a goniometer, with one arm positioned proximally at the middle of the metacarpal bone and the other arm placed distally at the middle of the first phalanx (Figure 2).

2.3.6. Fore Pastern Versus Ground Angle

A straight line was drawn from the fore pastern joint to the ground, extending distally and cranially. Another straight line was drawn horizontally at the leveled ground surface. The angle was measured by placing a protractor at the point where the line from the fore pastern intersects with the ground (Figure 2).

2.3.7. Hip Joint Angle

A straight line was drawn from the middle of the tuber coxae to the middle of the hip joint. Another straight line extended from the middle of the hip joint toward the middle of the stifle joint. A protractor was placed obliquely at the hip joint, and the angle was measured using a goniometer, positioning its disc at the center of the hip joint with the fixed arm pointing toward the tuber coxae and the movable arm toward the center of the femur bone (Figure 2).

2.3.8. Stifle Joint Angle

A straight line was drawn from the middle of the hip joint to the middle of the stifle joint. Another straight line was drawn from the middle of the stifle joint to the middle of the hock joint. A protractor was placed at the center of the stifle joint, and the angle was measured using a goniometer, with the disc positioned at the axis of the stifle joint. The fixed arm of the goniometer was aligned with the middle of the femur bone, while the other arm pointed towards the middle of the hock joint (Figure 2).

2.3.9. Hock Joint Angle

A straight line was drawn from the middle of the stifle joint to the middle of the hock joint. Another straight line was drawn from the middle of the hock joint to the middle of the hind fetlock joint. A protractor was positioned obliquely at the hock joint, and the angle was measured using a goniometer, with one arm placed at the proximal end of the first line and the other arm at the distal end of the second line (Figure 2).

2.3.10. Hind Fetlock Joint Angle

A straight line was drawn from the middle of the hock joint to the middle of the hind fetlock joint. Another straight line was drawn from the middle of the hind fetlock joint to the middle of the hind pastern joint. A protractor was placed at the hind fetlock joint, where both lines intersect, and the angle was measured using a goniometer, with one arm placed proximally at the first line and the second arm placed distally at the second line (Figure 2).

2.3.11. Hind Pastern Versus Ground Angle

A straight line was drawn from the middle of the hind pastern joint, extending cranially to the midpoint of the hoof and continuing toward the ground. A second line was drawn horizontally on the leveled ground surface. The angle between the two lines was measured by placing the protractor at the point where the lines intersect (Figure 2).

2.3.12. Girth Circumference and Body Length

The girth circumference was measured at the highest point of the withers and just behind the elbow and forelimb. This measurement was taken while the horse stood square at the level surface and at the moment of exhalation to ensure that additional lung volume was not included. Body length was measured from the point of the buttocks to the point of the shoulder.

2.4. Statistical Analysis

Age, girth circumference, BCS, body length, and weight are presented as mean and standard deviations (mean ± SD). All the studied angles are presented as box and violin plots in addition to mean ± SD. Data were tested for normality using the Shapiro–Wilk test. Horse joint angles and, age, body weight, and BCS were compared using Pearson correlation. Heatmaps were created based on the Pearson correlation coefficients (r) and p-values using R package ggplot2. All statistical analysis and data visualization were carried out using RStudio with R version 4.2.3 [26].

2.5. Ethical Approval

This study was approved by the Institutional Ethical Review Committee of the University of Veterinary and Animal Sciences Lahore, Pakistan (Approval reference no. DR/06).

3. Results

3.1. Body Measurements

The mean ± SD values for the age of the participating horses were 4.81 ± 0.67 years. The mean girth circumference and body length were 65.06 and 54.66 inches, respectively. The mean body weight was 350.42 kg, and the mean BCS was 4.90 out of 9 (Table 1).

Table 1. Mean and standard deviation (SD) values for age, girth circumference, body condition score (BCS), body length, and weight.

Measurements	Mean	SD
Age (years)	4.81	0.67
Girth Circumference (inch)	65.06	0.91
Body Length (inch)	54.66	1.02
Body Weight (kg)	350.42	15.56
BCS (/9)	4.90	0.81

3.2. Elbow, Knee, Fetlock, Stifle, and Hock Joint Angles

Mean ± SD values for elbow, knee, and fore fetlock joint angles were 123.02 ± 3.46°, 171.52 ± 2.39°, and 147.68 ± 5.11°, respectively. The mean ± SD values for stifle, hock, and hind fetlock joint angles were 128.62 ± 4.08°, 160.40 ± 3.89°, and 155.48 ± 2.68°, respectively (Figure 3).

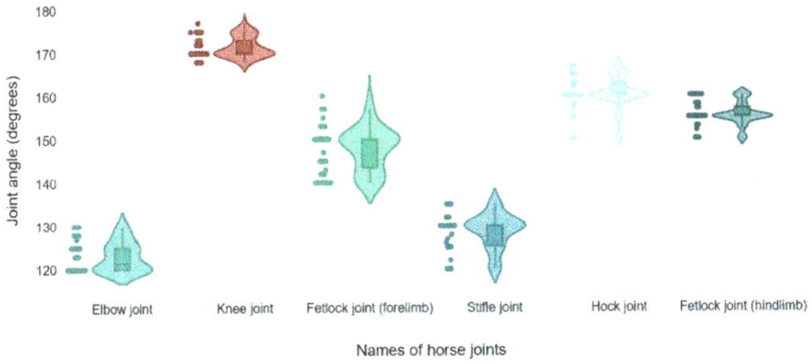

Figure 3. Box and violin plots for elbow, knee, fetlock (forelimb and hindlimb), stifle, and hock joint angles of the horses.

3.3. Shoulder, Pastern, and Hip Joint Angles

Mean ± SD values for shoulder and hip joint angles were 87.57 ± 2.94° and 85.82 ± 2.62°, respectively. The mean ± SD values for shoulder versus ground, fore pastern versus ground, and hind pastern versus ground angles were 58.38 ± 2.02°, 60.80 ± 3.53°, and 60.28 ± 4.01°, respectively (Figure 4).

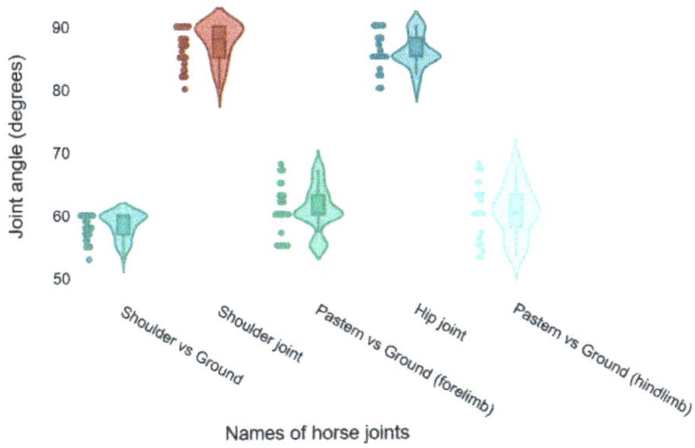

Figure 4. Box and violin plots for shoulder versus ground, shoulder, pastern versus ground (forelimb), pastern versus ground (hindlimb), and hip joint angles of the horses.

3.4. Correlation Between Horse Joint Angles, Age, Body Weight, and Body Condition Score

Non-significant correlations ($p > 0.05$) were found between horse joint angles and, age, body weight, and BCS. The Pearson correlation coefficients (r) and p-values between joint angles and, age, body weight, and BCS are presented (Figure 5).

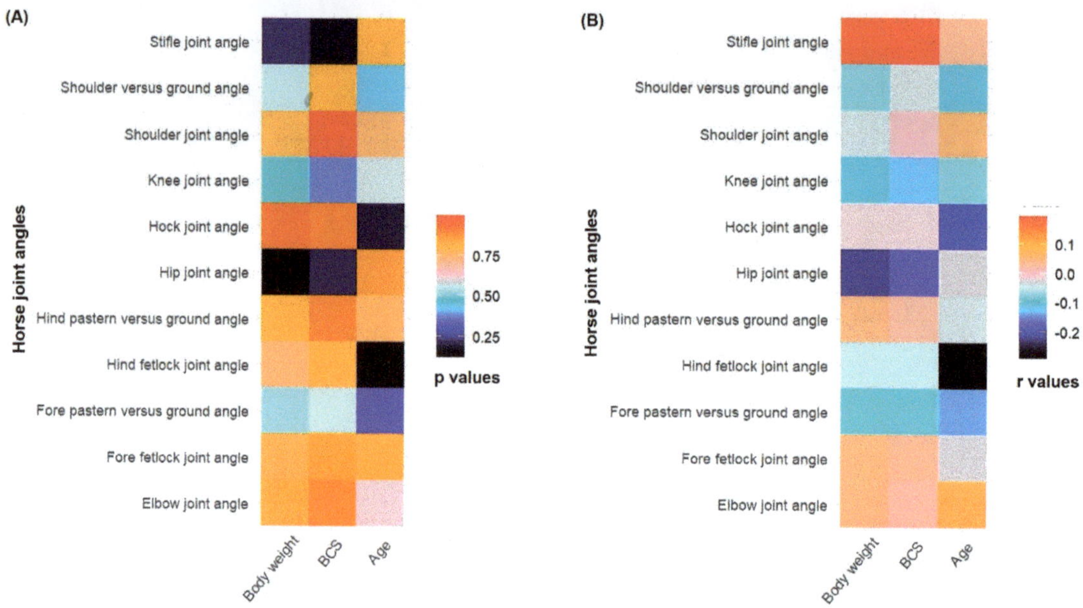

Figure 5. Heatmaps showing Pearson correlation between horse joint angles and age, body weight, and body condition score (BCS): (**A**) heatmap with respect to Pearson correlation p-values; (**B**) heatmap with respect to Pearson correlation coefficients (r values).

4. Discussion

This study aimed to evaluate and report normal limb joint angles and establish baseline measurements for joint conformation in the Morna breed of racehorses. The conformational assessment serves as a valuable indicator of soundness, aiding in the selection process to minimize the risk of lameness and identify horses with optimal racing potential [20]. Several methods for measuring joint angles are available, including goniometry, radiography, and advanced digital imaging techniques [22,24]. Among these, goniometry is important as it is simple, practical, cost-effective, less labor-intensive, repeatable, and reliable [22]. This method does not require sophisticated equipment and is widely used in conformational studies [22–24]. This is particularly significant in low-resource settings where access to advanced technology may be limited. We measured girth circumference, body length, weight, BCS, forelimbs, and hindlimb joint angles. The elbow joint and pastern versus ground angles in Morna racehorses show promising alignment with characteristics associated with optimal racing performance. However, further assessment and consideration of potential breed-specific variations would contribute to a more comprehensive understanding of their individual suitability and selection for racing and show jumping.

The correlations between horse joint angles and, age, body weight, and BCS were not significant. Currently, there are no studies directly comparing limb joint angles with age, body weight, or BCS. However, in mature horses, the angles of the metacarpophalangeal and distal interphalangeal joints remain relatively constant [27]. This may explain the non-significant correlation observed in Morna racehorses aged four to five years. Additionally, horse age is a significant predictor of radiographic changes in joints, with scores increasing by 0.2 points per year [28]. Interestingly, in beef cows, BCS was positively correlated with foot angle scores [29], likely due to the increased body weight associated with a higher BCS. Moreover, morphometric measurements, including limb lengths and joint angles, were associated with musculoskeletal disorders in jumping Thoroughbreds [30]. This underscores the need for further research on horse limb joint angles in relation to racing and associated skeletal malfunctions.

The shoulder joint angle of 87.56° in Morna racehorses is comparable to the findings in elite show jumping horses [31]. However, the recorded shoulder joint angle was not similar to previous studies conducted on jumping Thoroughbreds (99°) [14] and racing French trotters (116°) [16], respectively. This may be due to the varying methods used for measurement in different studies and the breed-specific differences. Moreover, differences in angle measurements may arise from errors in angular data due to skin displacement across anatomical landmarks, particularly in the proximal limb [32]. The elbow joint angle was 123.02° in Morna racehorses, consistent with the finding (124°) in racing French trotters [16]. While differing from the findings (138°) in jumping Thoroughbreds [14]. This difference may be attributed to breed-specific variations in conformation. The angle of the elbow joint is particularly important when analyzing the unique traits of French trotters in their racing performance, as even minor changes in the elbow angle can have a huge impact on the outcome of a race [16]. Moreover, the elbow joint of horses is crucial for stride length and overall movement fluidity, particularly in activities that require agility and speed [33]. A horse's potential increases with a lower elbow joint angle, and 122° is considered ideal elbow angle in French trotters [16]. In our study, the recorded elbow angle of 123.02° in Morna racehorses is close to the ideal value suggested by the previous study in French trotters [16]. Understanding these variations is crucial for breeders, trainers, and buyers to avoid potential limitations in athletic potential due to conformational defects.

The knee joint angle of 171.52° in Morna racehorses was similar to the angle observed in Swedish warmblood [31] and jumping Thoroughbred horses [14]. The optimal value of this angle should be 180° in racing French trotters [16]. The moderate variability, reflected in a standard deviation of 2.39° in Morna racehorses, underscores the need for individualized assessment when evaluating horses for races. The fore fetlock joint angle of 147.68° was seen in our study, whereas angles of 149° and 142° were seen in French trotters [16] and jumping Thoroughbreds [14], respectively. A higher fore fetlock joint angle increases the race potential of a horse, and 156° is considered an ideal value in racing French trotters [16].

The hip joint angle was 85.82° for Morna racehorses. Meanwhile, angles reported in studies conducted on Thoroughbred horses (88°) and racing French trotters (99°) were higher. Skin displacement at anatomical landmarks can lead to errors in angular measurements [32]. Interesting interbreed variations may originate from different degrees of skin movement rather than true variability in skeletal mobility. This may explain the slight angle difference between our and previous studies measuring joint angles. A recent study suggests that a horse's potential for racing performance increases with the decrease in hip angle, and an ideal angle value is 90° [16].

The stifle angle in our study was 128.62°, which is greater than the angle (114°) reported in jumping Thoroughbred horses [14]. A higher stifle angle increases the potential of a racehorse [16]. The hock angle of 160.40° in Morna racehorses is comparable to Swedish warmblood horses [31]. However, our results are inconsistent with the findings in jumping Thoroughbred horses (148°) and French trotters (142°) respectively [14,16,20]. This may be because of interbreed differences in joint angles. Furthermore, the ideal value of hock angle is 142° in French trotters, and a decrease in stifle angle increases the race potential of a horse [16].

The hind fetlock angle was 155.48°, consistent with the findings (157°) in jumping Thoroughbred horses [14]. However, our results are inconsistent with the findings (142°) in French trotters [16]. Furthermore, it is reported that the optimal value of the hind fetlock angle is 142°, and increasing the hind fetlock angle increases the race potential of a horse [16]. In summary, the hip angle, stifle angle, and the angle of the pastern in relation to the ground in Morna racehorses exhibit promising alignment with characteristics associated with optimal racing performance. However, the angle of hock and fetlock differ from those found in previous studies of jumping Thoroughbreds and French trotters. It is important to note that we had not calculated the sample size or conducted a power analysis, which may present a potential limitation of this study. We recommend conducting further research to explore the conformation of this specific breed using advanced techniques and equipment, such as modern imaging methods and artificial intelligence software, for example, Quintic Biomechanics video analysis software. Special emphasis should be placed on measuring the lengths of limb segments and correlating these measurements with the angles. This approach may provide valuable insights into individual variations within the breed.

5. Conclusions

Our study offered valuable insights into the conformational features of Morna racehorses. The joint angles and body measurements we observed provide a foundation for creating a database for selecting Morna racehorses, considering their unique traits. Breeders, trainers, and buyers need to understand the breed-specific variations discussed in our study to make informed decisions about the athletic potential and performance capabilities of individual Morna racehorses. Additional research should focus on examining the conformation of the Morna breed, particularly by measuring the lengths of limb segments and correlating them with joint angles. This investigation may reveal further insights into individual variations within the breed.

Author Contributions: I.A.: original draft preparation, writing—reviewing and editing, data collection, data curation, methodology, investigation, validation. S.I.: writing—reviewing and editing, data curation, methodology, investigation, validation, supervision. M.M.U.: writing—reviewing and editing, investigation, supervision. A.S.: writing—reviewing and editing, investigation, supervision. I.U.K.: writing—reviewing and editing, validation. M.Z.: writing—reviewing and editing, validation. S.S.U.H.B.: writing—original draft, writing—reviewing and editing, software, data curation, visualization, and formal data analysis. All authors participated in the research, contributed to the preparation, and approved the final version of the manuscript. All authors have read and agreed to the published version of the manuscript.

Funding: This research received no external funding.

Institutional Review Board Statement: This study was approved by the Institutional Ethical Review Committee of the University of Veterinary and Animal Sciences Lahore, Pakistan (Approval reference no. DR/06).

Informed Consent Statement: Informed consent was obtained from all subjects involved in the study.

Data Availability Statement: The supporting raw data of this research will be made available upon request.

Conflicts of Interest: The authors declare no conflicts of interest.

References

1. Klecel, W.; Martyniuk, E. From the eurasian steppes to the roman circuses: A review of early development of horse breeding and management. *Animals* **2021**, *11*, 1859. [CrossRef]
2. Steel, C.; Morrice-West, A. Veterinary Aspects of Training, Conditioning, and Racing Thoroughbred Racehorses. In *Equine Sports Medicine and Surgery*; Elsevier: Amsterdam, The Netherlands, 2024; pp. 1131–1168.

3. McIlwraith, C.W.; Anderson, T.M.; Douay, P. The role of conformation in musculoskeletal problems in the racing Thoroughbred. *Equine Vet. J.* **2004**, *36*, 571–575. [CrossRef]
4. Bukhari, S.S.U.H.; McElligott, A.G.; Parkes, R.S.V. Quantifying the Impact of Mounted Load Carrying on Equids: A Review. *Animals* **2021**, *11*, 1333. [CrossRef]
5. Bukhari, S.S.U.H.; Parkes, R.S.V. Assessing the impact of draught load pulling on welfare in equids. *Front. Vet. Sci.* **2023**, *10*, 1214015. [CrossRef] [PubMed]
6. Ripollés-Lobo, M.; Perdomo-González, D.I.; Valera, M.; Gómez, M.D. Conformational Defects in the Limbs of Menorca Purebred Horses and Their Relationship to Functionality. *Animals* **2024**, *14*, 1071. [CrossRef] [PubMed]
7. Domańska-Kruppa, N.; Wierzbicka, M.; Stefanik, E. Advances in the Clinical Diagnostics to Equine Back Pain: A Review of Imaging and Functional Modalities. *Animals* **2024**, *14*, 698. [CrossRef] [PubMed]
8. Robert, C.; Valette, J.-P.; Denoix, J.-M. Longitudinal development of equine forelimb conformation from birth to weaning in three different horse breeds. *Vet. J.* **2013**, *198*, e75–e80. [CrossRef] [PubMed]
9. Van Weeren, P.R.; Crevier-Denoix, N. Equine conformation: Clues to performance and soundness? *Equine Vet. J.* **2006**, *38*, 591–596. [CrossRef]
10. Crevier-Denoix, N.; Camus, M.; Falala, S.; Martino, J.; Pauchard, M.; Ravary-Plumioen, B.; Desquilbet, L.; Denoix, J.-M.; Chateau, H.; Pourcelot, P. Comparison of the loading of leading and trailing forelimbs in horses at landing. *Equine Vet. J.* **2014**, *46*, 38. [CrossRef]
11. Boswell, R.P.; Mitchell, R.D.; Dyson, S.J. Lameness in the show hunter and show jumper. In *Diagnosis and Management of Lameness in the Horse*; Elsevier: Amsterdam, The Netherlands, 2011; pp. 965–975.
12. Schlueter, A.E.; Orth, M.W. Equine osteoarthritis: A brief review of the disease and its causes. *Equine Comp. Exerc. Physiol.* **2004**, *1*, 221–231. [CrossRef]
13. Yildirim, I.I.G.; Erden, H. Conformational characteristics in Arabian and Thoroughbred horses. *Anim. Heal. Prod. Hyg.* **2023**, *12*, 27–35. [CrossRef]
14. Senna, N.A.; Mostafa, M.B.; Abu-Seida, A.M.; Elemmawy, Y.M. Evaluation of limb conformation in jumping thoroughbred horses. *Asian J. Anim. Sci.* **2015**, *9*, 208–216. [CrossRef]
15. Agass, R.F.; Wilson, A.M.; Weller, R.; Pfau, T. The Relationship Between Foot Conformation, Foot Placement and Motion Symmetry in the Equine Hindlimb. *Equine Vet. J.* **2014**, *46*, 19–20. [CrossRef]
16. Górniak, W.; Walkowicz, E.; Soroko, M.; Korczyński, M. Impact of the individual characteristics of French trotters on their racing performance. *Turkish J. Vet. Anim. Sci.* **2020**, *44*, 110–117. [CrossRef]
17. Labuschagne, W.; Rogers, C.W.; Gee, E.K.; Bolwell, C.F. A Cross-Sectional Survey of Forelimb Hoof Conformation and the Prevalence of Flat Feet in a Cohort of Thoroughbred Racehorses in New Zealand. *J. Equine Vet. Sci.* **2017**, *51*, 1–7. [CrossRef]
18. McIlwraith, C.W.; Anderson, T.M.; Sanschi, E.M. Conformation and musculoskeletal problems in the racehorse. *Clin. Tech. Equine Pract.* **2003**, *2*, 339–347. [CrossRef]
19. Holmes, T.Q.; Brown, A.F. Champing at the bit for improvements: A review of equine welfare in equestrian sports in the United Kingdom. *Animals* **2022**, *12*, 1186. [CrossRef] [PubMed]
20. Mostafa, M.B.; Senna, N.A.; Abu-Seida, A.M.; Elemmawy, Y.M. Evaluation of abnormal limb conformation in jumping Thoroughbred horses. *J. Hell. Vet. Med. Soc.* **2019**, *70*, 1533–1540. [CrossRef]
21. Gmel, A.I.; Druml, T.; Portele, K.; von Niederhäusern, R.; Neuditschko, M. Repeatability, reproducibility and consistency of horse shape data and its association with linearly described conformation traits in Franches-Montagnes stallions. *PLoS ONE* **2018**, *13*, e0202931. [CrossRef] [PubMed]
22. Hancock, G.E.; Hepworth, T.; Wembridge, K. Accuracy and reliability of knee goniometry methods. *J. Exp. Orthop.* **2018**, *5*, 46. [CrossRef] [PubMed]
23. Jacklin, B.D.; Hanousek, K.; Gillespie, S.; Liedtke, A.; Tucker, R.; Fiske-Jackson, A.; Smith, R.K. Validation of a novel clinical tool for monitoring distal limb stiffness. *Front. Vet. Sci.* **2023**, *10*, 127103. [CrossRef] [PubMed]
24. Tuan, C.C.; Wu, Y.C.; Yeh, W.L.; Wang, C.C.; Lu, C.H.; Wang, S.W.; Yang, J.; Lee, T.F.; Kao, H.K. Development of Joint Activity Angle Measurement and Cloud Data Storage System. *Sensors* **2022**, *22*, 4684. [CrossRef]
25. Jensen, R.B.; Rockhold, L.L.; Tauson, A.H. Weight estimation and hormone concentrations related to body condition in Icelandic and Warmblood horses: A field study. *Acta Vet. Scand.* **2019**, *61*, 63. [CrossRef]
26. RStudio. Available online: https://www.rstudio.com/ (accessed on 11 November 2024).
27. Addis, P.R.; Lawson, S.E.M. The role of tendon stiffness in development of equine locomotion with age. *Equine Vet. J.* **2010**, *42*, 556–560. [CrossRef]
28. Ramos, S.; Pinto, A.; Cardoso, M.; Alexandre, N.; Bettencourt, E.; Monteiro, S.; Gama, L.T. Prevalence of Radiographic Signs of Osteoarthritis in Lusitano Purebred Horses. *J. Equine Vet. Sci.* **2020**, *94*, 103196. [CrossRef] [PubMed]
29. Banwarth, M.R.; DeAtley, K.; Doyle, S. PSI-38 Relationship between body condition score, claw set, and foot angle in Angus and Red Angus beef cows grazing native annual-grass rangeland in northern California. *J. Anim. Sci.* **2020**, *98*, 474. [CrossRef]

30. Mostafa, M.B.; Elemmawy, Y.M. Relationships between morphometric measurements and musculoskeletal disorders in jumping thoroughbred horses. *J. Equine Sci.* **2020**, *31*, 23–27. [CrossRef]
31. Holmström, M.; Magnusson, L.E.; Philipsson, J. Variation in conformation of Swedish warmblood horses and conformational characteristics of élite sport horses. *Equine Vet. J.* **1990**, *22*, 186–193. [CrossRef] [PubMed]
32. Van Weeren, P.R.; Van Den Bogert, A.J.; Barneveld, A. A quantitative analysis of skin displacement in the trotting horse. *Equine Vet. J.* **1990**, *22*, 101–109. [CrossRef]
33. Harrison, S.M.; Whitton, R.C.; King, M.; Haussler, K.K.; Kawcak, C.E.; Stover, S.M.; Pandy, M.G. Forelimb muscle activity during equine locomotion. *J. Exp. Biol.* **2012**, *215*, 2980–2991. [CrossRef]

veterinary sciences

MDPI

Article

Evaluation of Serum Lipids, Biochemical Parameters, Selected Antioxidant Elements and Oxidative Stress Profiles in Late Pregnant Jennies with Hyperlipemia

Qingze Meng †, Yang Shao †, Wei Li, Jia Lu, Xinyue Wang and Liang Deng *

Department of Animal Genetics, Breeding and Reproduction, College of Animal Science and Veterinary Medicine, Shenyang Agricultural University, Shenyang 110866, China; 15502462577@163.com (Q.M.); 2022220555@stu.syau.edu.cn (Y.S.); 2023220582@stu.syau.edu.cn (W.L.); 2023240732@stu.syau.edu.cn (J.L.); 2022240704@stu.syau.edu.cn (X.W.)

* Correspondence: ldeng@syau.edu.cn

† These authors contributed equally to this work.

Simple Summary: Hyperlipemia is common in donkeys and has a reported prevalence of 3–5% in the field population. In this study, the serum levels of lipids, biochemical parameters, selected antioxidant elements and oxidative stress parameters in late pregnant jennies (female donkeys) with and without hyperlipemia were evaluated and compared. Key findings showed that serum had elevated levels of lipids, including triglycerides, total cholesterol and low-density lipoprotein cholesterol in late pregnant jennies with hyperlipemia. In contrast, serum levels of selenium and antioxidant capacity diminished. This study revealed the features of oxidative stress in late pregnant hyperlipemic jennies underscored the importance of early diagnosis and intervention for hyperlipidemia in pregnant jennies. It has potential implications for improving clinical diagnostic methods and treatment strategies.

Abstract: Donkeys are particularly at risk of hyperlipemia. Hyperlipemia is a metabolic disease caused by the mobilization of fatty acids from adipose tissue, which often impacts pregnant and lactating jennies (female donkeys) during periods of negative energy balance. This study aimed to evaluate the levels of lipids, biochemical parameters, selected antioxidant elements and oxidative stress parameters in late pregnant jennies affected by hyperlipemia. Compared with the healthy jennies, the hyperlipemic animals exhibited significantly elevated levels of triglycerides (TGs), total cholesterol (T-CHO) and low-density lipoprotein cholesterol ($p < 0.05$), coupled with reduced levels of high-density lipoprotein cholesterol and albumin (ALB) ($p < 0.05$). The serum levels of biochemical parameters related to liver function, such as aspartate aminotransferase, alanine aminotransferase, alkaline phosphatase (AKP) and cholinesterase (CHE), showed a significant increase in the hyperlipemia group compared to the healthy group ($p < 0.05$). The serum level of selenium was significantly lower ($p < 0.05$) and positively correlated with TGs ($r = 0.85$) and ALB ($r = 0.73$) in the hyperlipemia group. The hyperlipemic jennies showed diminished serum levels of antioxidant capacity and increased levels of malondialdehyde (MDA). The area under the curve values for T-CHO, ALB, AKP, CHE, total superoxide dismutase, glutathione and MDA were relatively high. Thus, our findings reflect metabolic disorders, liver dysfunction and oxidative stress in late pregnant hyperlipemic jennies, providing a basis for the improvement of clinical diagnostic methods and early prevention and control of hyperlipemia in jennies.

Keywords: donkey; hyperlipemia; oxidative stress; pregnancy; trace element

1. Introduction

Donkeys (*Equus africanus asinus*) have been used by humans for pack and draught work. Nowadays, the new and evolving role of donkeys in milk, meat and hide production and animal-assisted therapy is a renewed interest to human life [1]. Donkeys have a long reproductive cycle, where the gestation length lasts for a mean value of 374 days [2]. Therefore, successful pregnancies are essential for maintaining the reproductive efficiency of donkeys. However, pregnancy is a very dynamic physiological period that leads to an increase in metabolic demand [3]. In the final trimester of the pregnancy period, the jenny (female donkey) has a greater energy requirement as the fetus develops rapidly, leading to the negative energy balance in the maternal body [4].

Dyslipidemia is more frequent in donkeys than in other equids. Donkeys are particularly at risk of hyperlipemia, with a prevalence of 3–5% in the general population [5]. The higher incidence of hyperlipemia in donkeys is likely related to their efficiency to store lipids and rapid ability to mobilize fat stores [6]. Hyperlipemia is a metabolic disease caused by the mobilization of fatty acids from adipose tissue and is typically associated with periods of negative energy balance and physiologic disturbances of glucose and lipid homeostasis, which frequently affects late pregnant and lactating donkeys [7,8]. In late pregnancy, estrogen stimulates the liver to produce very low-density lipoprotein (VLDL), reducing the clearance of triglycerides (TGs) by lipoprotein lipase in the liver and adipose tissue. It leads to an increase in plasma triglyceride concentration to ensure sufficient energy substrates for normal fetal development [6]. The donkeys with hyperlipemia commonly showed clinical signs of dullness, depression, anorexia, weakness, emaciation, rough hair coat, continuous weight loss, lethargy and poor body condition [9]. They are invariably difficult to manage and mortality rates in the range of 60–80% are frequently reported, which causes significant economic losses in the global donkey industry [10].

However, hyperlipemia usually has no obvious clinical manifestations in the early stage, which may not be noticeable early on to the owners. Hyperlipemia is diagnosed by the routine measurement of serum TG concentrations in sick or at-risk donkeys [9]. Donkeys have lower LDL and cholesterol, slightly higher high-density lipoprotein (HDL), but similar VLDL concentrations compared with horses [6]. This different lipoprotein profile in donkeys could potentially contribute to their higher risk of hyperlipemia.

Oxidative damage can accelerate the pathogenic progress of hyperlipemia and its complications. In the early evolution of hyperlipidemia, oxidative stress and lipoprotein oxidation played an important role [11]. Lipids, in particular, the polyunsaturated fatty acyl chains contained in phospholipids, represent a main target of reactive oxygen species (ROS) attack [12]. LDLs represent a prominent target for an oxidative reaction [11]. An adequate antioxidant supply may help prevent the course of hyperlipemia [13]. Some trace elements and vitamins were reported to have an antioxidant capacity and be vital to maintaining normal lipid metabolism in humans and animals. The functions of selenium (Se) in the organism are mainly connected with its antioxidant properties, as it is an essential part of important antioxidant enzymes [14]. Zinc (Zn) is a component of the oxidant defense system and associated with oxidative stress in relation to its deficiency [15]. Vitamin E (VE) is the major lipophilic radical-scavenging antioxidant in vivo and protects the body from the oxidative stress mediated by active oxygen [16]. Furthermore, a meta-analysis showed that serum Se was associated with hyperlipidemia in humans [17]. VE possess anti-hyperlipidemic properties in various types of animals [18].

To date, to the best of our knowledge, little is known about the changes in the serum levels of lipids, biochemical parameters, antioxidant elements and oxidative stress parameters in jennies with and without hyperlipemia in late pregnancy; however, more detailed information on these parameters during late pregnancy in jennies is important to ensure proper diagnosis, care and disease treatment. Therefore, we hypothesized that some serum antioxidant elements might play a vital role in regulating lipid metabolism and oxidative stress in late pregnant jennies with hyperlipemia. The objective of the present study was to evaluate and compare the levels of lipids, biochemical parameters, selected antioxidant elements and oxidative stress parameters between late pregnant jennies with and without hyperlipemia.

2. Materials and Methods

2.1. Animal Welfare Statement

All experiments were approved by the Animal Care and Use Committee of Shenyang Agricultural University (Approval no. 202306017).

2.2. Animals

The study was conducted at a donkey breeding farm in Liaoning Province, China. Forty Chinese Liaoxi jennies in late pregnancy were included in this study. Among these jennies, twenty animals were clinically diagnosed with hyperlipemia based on the following criteria: (1) developed depression, anorexia, weight loss, a rough coat and lethargy; (2) body condition scores (BCSs) ranged from 1 to 2 (1 = poor, 2 = moderate, 3 = ideal, 4 = fat and 5 = obese); (3) serum was usually turbid and white (milky); and (4) the serum TG concentrations were over 2.8 mmol/L, according to the guidelines [9,19]. The other twenty jennies in late pregnancy were evaluated clinically following The Donkey Sanctuary recommended standards, including demeanor and behavior, body condition, temperature, pulse rate, respiratory rate, mucous membrane color, abdominal auscultation, examination of the oral cavity, ocular examination, rectal examination, peritoneal tap, ultrasound and blood examination [20]. They were confirmed to be healthy throughout the investigation. The comparison of age, parity, body weight, BCS and gestation length in hyperlipemic and healthy jennies is shown in Table 1.

Table 1. Comparison of age, parity, body weight, BCS and gestation length in hyperlipemic (n = 20) and healthy (n = 20) jennies.

Parameters	Hyperlipemia	Healthy	p Value
Age	6.1± 0.2	5.8 ± 0.2	0.403
Parity	1.9 ± 0.1	1.8 ± 0.2	0.687
Body weight (kg)	209.4 ± 4.1 ^{b}	247.5 ± 3.8 ^{a}	0.000
BCS	1.9 ± 0.1 ^{b}	3.1 ± 0.1 ^{a}	0.000
Gestation length (days)	367.2 ± 1.6	362.7 ± 1.5	0.392

Abbreviations: BCS, body condition score. ^{a,b} In the same row, values with different letters in superscript differ significantly (p < 0.05).

Jennies were fed with dry corn stover, along with a commercial concentrate feed [21], according to the nutrient requirements stated by the National Research Council recommendations [22]. They were housed in individual boxes (3 m × 3 m) 10–15 days before the presumptive delivery until 15 days after foaling. Measures were implemented to minimize stress throughout the study.

2.3. Blood Sampling

Blood samples were obtained from the jugular vein between 7:30 and 8:30 a.m. to avoid alterations related to diurnal variations [23]. For the determination of serum parameters, blood samples were collected into 10 mL vacuum tubes (BD Vacutainer, NJ, USA) without anticoagulant. The blood samples were then centrifuged at 3000× g for 10 min at 4 °C to separate the serum. Subsequently, the serum samples were stored at −80 °C until further analysis.

2.4. Serum Parameter Analysis

Serum levels of TGs, total cholesterol (T-CHO), high-density lipoprotein cholesterol (HDL-C), low-density lipoprotein cholesterol (LDL-C), total protein (TP), albumin (ALB), aspartate aminotransferase (AST), alanine aminotransferase (ALT), alkaline phosphatase (AKP), cholinesterase (CHE), total antioxidant capacity (T-AOC), glutathione (GSH), total superoxide dismutase (T-SOD) and malondialdehyde (MDA) were measured using a microplate reader (GENios Plus, Tecan, Männedorf, Switzerland) following standard methods with commercially available kits (Nanjing Jiancheng Bioengineering Institute,

Nanjing, China). The serum levels of Se and Zn were determined by the established procedures of atomic fluorescence spectrometry (HGF-V, Beijing Haiguang Instrument Co., Ltd., Beijing, China) and atomic absorption spectrometry (model Z-2000, Hitachi, Tokyo, Japan), respectively. High-performance liquid chromatography was used to determine the content of VE in the serum by Agilent High-Performance Liquid Chromatograph 1260 Series (Agilent Technologies, Santa Clara, CA, USA).

2.5. Statistical Analysis

Data were analyzed with the statistical software program SPSS for Windows, version 22.0 (SPSS Inc., Chicago, IL, USA). Data were assessed for normality by the Shapiro–Wilk test, which showed that the variables were normally distributed. Simple descriptive measures, such as mean, standard error (SE) and confidence interval (CI) values, of all parameters were estimated. According to the Youden index, receiver operating characteristic (ROC) curve analysis was conducted to estimate optimal values of cutoff, as well as to maximize sensitivity and specificity. An independent-sample Student t test was used to determine the statistical significance of differences in serum parameters between hyperlipemic and healthy jennies. Pearson correlation was performed on the serum parameters of hyperlipemic and healthy jennies. All values were presented as mean ± SE. Significant differences were considered in terms of the associated p-value relative to $p < 0.05$ and $p < 0.01$.

3. Results

3.1. Analysis of Serum Lipids

The serum levels of TGs, T-CHO and LDL-C showed a significant increase in the hyperlipemia group compared to the healthy group ($p < 0.05$). The serum level of HDL-C was significant lower in the hyperlipemia group than that in the healthy group ($p = 0.002$) (Table 2).

Table 2. Comparison of serum lipids for hyperlipemic ($n = 20$) and healthy ($n = 20$) jennies.

Parameters		Hyperlipemia	Healthy	p Value
TG (mmol/L)	Mean ± SE	21.35 ± 1.44 [a]	1.84 ± 0.14 [b]	0.000
	95% CI	18.31~24.39	1.55~2.12	
T-CHO (mmol/L)	Mean ± SE	10.75 ± 1.14 [a]	3.50 ± 0.15 [b]	0.000
	95% CI	8.36~13.15	3.20~3.79	
HDL-C (mmol/L)	Mean ± SE	1.85 ± 0.22 [b]	2.86 ± 0.16 [a]	0.002
	95% CI	1.37~2.32	2.53~3.18	
LDL-C (mmol/L)	Mean ± SE	3.46 ± 0.37 [a]	1.96 ± 0.15 [b]	0.001
	95% CI	2.69~4.23	1.66~2.26	

Abbreviations: TG, triglyceride; T-CHO, total cholesterol; HDL-C, high-density lipoprotein cholesterol; LDL-C, low-density lipoprotein cholesterol; SE, standard error; CI: confidence interval. [a,b] In the same row, values with different letters in superscript differ significantly ($p < 0.05$).

Analytic results of the ROC curve are shown in Figure 1. The area under the curve (AUC) values of serum TG and T-CHO levels in the hyperlipemia and healthy groups were 1.000 and 0.972, with the optimum cutoff point calculated on the basis of maximum sensitivity and specificity being 8.19 mmol/L and 5.88 mmol/L, respectively. The AUCs of serum HDL-C and LDL-C were 0.719 and 0.600, with the optimum cutoff point being 2.40 mmol/L and 2.65 mmol/L, respectively. The AUC of HDL-C/LDL-C was 0.835, with the optimum cutoff point being 0.97 (Figure 1a).

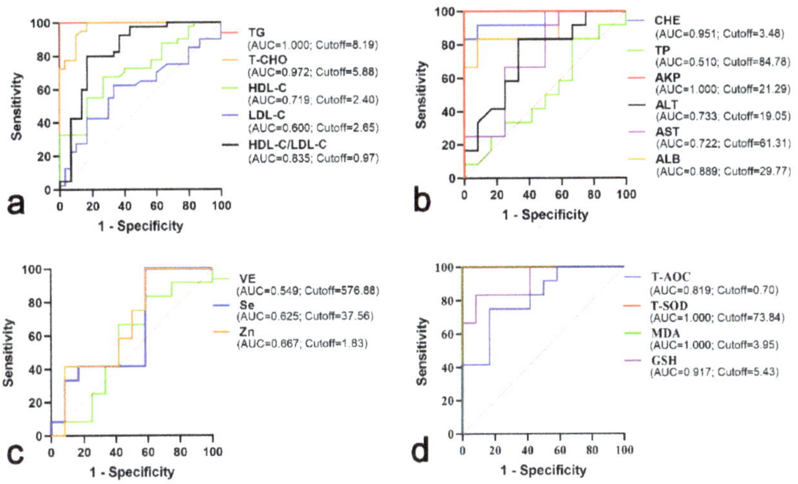

Figure 1. ROC of serum parameters in hyperlipemic and healthy jennies. (**a**) ROC of serum lipids; (**b**) ROC of serum biochemical parameters related to liver function; (**c**) ROC of serum selenium, zinc and vitamin E; (**d**) ROC of serum oxidative stress parameters. Abbreviations: ROC, receiver operating characteristic; TG, triglyceride; T-CHO, total cholesterol; HDL-C, high-density lipoprotein cholesterol; LDL-C, low-density lipoprotein cholesterol; TP, total protein; ALB, albumin; AST, aspartate aminotransferase; ALT, alanine aminotransferase; AKP, alkaline phosphatase; CHE, cholinesterase; T-AOC, total antioxidant capacity; T-SOD, total superoxide dismutase; MDA, malondialdehyde; GSH, glutathione; VE, vitamin E; Se, selenium; Zn, zinc.

3.2. Analysis of Serum Biochemical Parameters Related to Liver Function

The descriptive statistics and differences in serum levels of biochemical parameters related to liver function between the hyperlipemic and healthy jennies are summarized in Table 3. The serum level of ALB in the hyperlipemic jennies was significantly lower than in the healthy animals ($p = 0.001$), and the TP level was not found to be statistically different between the two groups ($p = 0.490$). The serum levels of AST, ALT, AKP and CHE showed a significant increase in the hyperlipemia group compared to the healthy group ($p < 0.05$).

Table 3. Comparison of serum biochemical parameters related to liver function between hyperlipemic ($n = 20$) and healthy ($n = 20$) jennies.

Parameters		Hyperlipemia	Healthy	p Value
TP (g/L)	Mean ± SE	77.80 ± 2.79	75.55 ± 1.61	0.490
	95% CI	71.97~83.64	72.16~78.94	
ALB (g/L)	Mean ± SE	25.84 ± 0.96 ^{b}	31.98 ± 1.22 ^{a}	0.001
	95% CI	23.72~27.96	29.29~34.68	
AST (U/L)	Mean ± SE	68.51 ± 20.41 ^{a}	37.33 ± 3.87 ^{b}	0.019
	95% CI	21.45~115.56	29.44~45.21	
ALT (U/L)	Mean ± SE	22.41 ± 3.24 ^{a}	13.08 ± 1.21 ^{b}	0.013
	95% CI	15.61~29.21	10.65~15.51	
AKP (King unit/100 mL)	Mean ± SE	66.70 ± 6.45 ^{a}	15.08 ± 0.73 ^{b}	0.000
	95% CI	53.20~80.20	13.55~16.61	
CHE (U/mL)	Mean ± SE	4.15 ± 0.19 ^{a}	2.11 ± 0.19 ^{b}	0.000
	95% CI	3.75~4.55	1.71~2.52	

Abbreviations: TP, total protein; ALB, albumin; AST, aspartate aminotransferase; ALT, alanine aminotransferase; AKP, alkaline phosphatase; CHE, cholinesterase; SE, standard error; CI: confidence interval. ^{a,b} In the same row, values with different letters in superscript differ significantly ($p < 0.05$).

The AUCs of serum TP and ALB levels in the hyperlipemia and healthy groups were 0.510 and 0.889, with the optimum cutoff points being 84.78 g/L and 29.77 g/L, respectively. The AUCs of serum levels of AST, ALT, AKP and CHE were 0.722, 0.733, 1.000 and 0.951, with the optimum cutoff points being 61.31 U/L, 19.05 U/L, 21.29 King unit/100 mL and 3.48 U/mL, respectively (Figure 1b).

3.3. Analysis of Serum Selenium, Zinc and Vitamin E

As shown in Table 4, the serum level of Se is lower in the hyperlipemia group compared with the healthy group ($p = 0.031$). There was no significant difference in serum levels of Zn and VE between the two groups ($p > 0.05$).

Table 4. Comparison of serum selenium, zinc and vitamin E in hyperlipemic ($n = 20$) and healthy ($n = 20$) jennies.

Parameters		Hyperlipemia	Healthy	p Value
Se (μg/L)	Mean ± SE	42.19 ± 4.65 [b]	55.51 ± 3.12 [a]	0.031
	95% CI	31.67~52.71	48.32~62.71	
Zn (mg/L)	Mean ± SE	1.60 ± 0.13	1.45 ± 0.09	0.345
	95% CI	1.29~1.90	1.24~1.65	
VE (nmol/L)	Mean ± SE	560.80 ± 43.07	552.58 ± 43.33	0.894
	95% CI	463.36~658.24	454.55~650.61	

Abbreviations: Zn, zinc; Se, selenium; VE, vitamin E; SE, standard error; CI: confidence interval. [a,b] In the same row, values with different letters in superscript differ significantly ($p < 0.05$).

The AUC of serum Se in the hyperlipemia and healthy groups was 0.625, with the optimum cutoff point being 37.56 μg/L. The AUCs of serum Zn and VE were 0.667 and 0.549, with the optimum cutoff points being 1.83 mg/L and 576.88 nmol/L, respectively (Figure 1c).

3.4. Analysis of Serum Oxidative Stress Parameters

Comparisons of the serum levels of oxidative stress parameters in the hyperlipemia and healthy groups are shown in Table 5. The serum level of MDA in the hyperlipemia group was higher than that of the healthy group ($p = 0.000$), and the levels of T-AOC, GSH and T-SOD were lower than those in healthy group ($p < 0.05$).

Table 5. Comparison of serum oxidative stress parameters for hyperlipemic ($n = 20$) and healthy ($n = 20$) jennies.

Parameters		Hyperlipemia	Healthy	p Value
T-AOC (mmol/L)	Mean ± SE	0.67 ± 0.01 [b]	0.71 ± 0.01 [a]	0.004
	95% CI	0.65~0.68	0.69~0.72	
GSH (μmol/L)	Mean ± SE	2.87 ± 0.35 [b]	9.92 ± 1.06 [a]	0.000
	95% CI	2.15~3.59	7.70~12.13	
T-SOD (U/mL)	Mean ± SE	70.80 ± 0.15 [b]	76.85 ± 0.18 [a]	0.000
	95% CI	70.49~71.12	76.47~77.23	
MDA (nmol/mL)	Mean ± SE	6.34 ± 0.47 [a]	2.78 ± 0.09 [b]	0.000
	95% CI	5.36~7.32	2.59~2.96	

Abbreviations: T-AOC, total antioxidant capacity; GSH, glutathione; T-SOD, total superoxide dismutase; MDA, malondialdehyde; SE, standard error; CI: confidence interval. [a,b] In the same row, values with different letters in superscript differ significantly ($p < 0.05$).

The AUCs of serum T-AOC and GSH concentrations in the hyperlipemia and healthy groups were 0.819 and 0.917, with the optimum cutoff points being 0.70 mmol/L and 5.43 μmol/L, respectively. The AUCs of serum T-SOD and MDA were both 1.000, with the optimum cutoff points being 73.84 U/mL and 3.95 nmol/mL, respectively (Figure 1d).

3.5. Analysis of Correlation Among Serum Parameters

The correlations among serum lipids, biochemical parameters, selected antioxidant elements and oxidative stress parameters of hyperlipemic and healthy jennies are shown in Figure 2. The serum level of Se was positively associated with TGs (r = 0.85, p = 0.005) and ALB (r = 0.73, p = 0.014) in the hyperlipemia group. The serum level of LDL-C was negatively associated with GSH (r = 0.71, p = 0.033) and AKP (r = 0.74, p = 0.009). The serum level of Zn was positively associated with T-AOC (r = 0.87, p = 0.005) (Figure 2). In healthy jennies, the serum level of T-CHO was positively associated with HDL-C (r = 0.74, p = 0.006), LDL-C (r = 0.89, p = 0.008) and AKP (r = 0.61, p = 0.037). MDA was negatively associated with ALT (r = 0.84, p = 0.004) and Zn (r = 0.78, p = 0.038) (Figure 2).

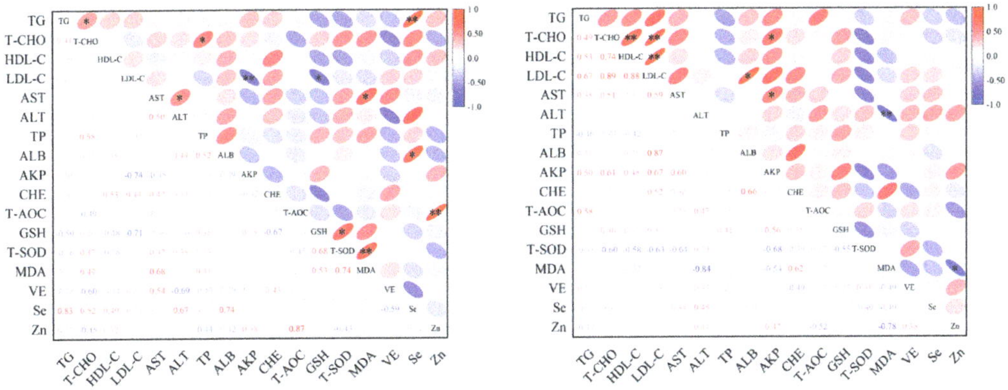

Figure 2. Correlation heat map of serum lipids, biochemical parameters, selected antioxidant elements and oxidative stress parameters of hyperlipemic and healthy jennies. *, $p < 0.05$; **, $p < 0.01$. Abbreviations: TG, triglyceride; T-CHO, total cholesterol; HDL-C, high-density lipoprotein cholesterol; LDL-C, low-density lipoprotein cholesterol; AST, aspartate aminotransferase; ALT, alanine aminotransferase; TP, total protein; ALB, albumin; AKP, alkaline phosphatase; CHE, cholinesterase; T-AOC, total antioxidant capacity; GSH, glutathione; T-SOD, total superoxide dismutase; MDA, malondialdehyde; VE, vitamin E; Se, selenium; Zn, zinc.

4. Discussion

Late pregnancy is the most challenging and critical period of breeding jennies. The digestive tract capacity decreases as the pregnancy progresses, leading to an inability of being able to satisfy the energy requirements for the jenny and the foal. Thus, it is a major predisposing factor for hyperlipemia in the late pregnancy stage of jennies [24]. In this study, we compared the levels of lipids, biochemical parameters, selected antioxidant elements and oxidative stress parameters for jennies with and without hyperlipemia. Our results provide a deep understanding of evaluating the features of hyperlipemia in jennies.

We observed serum lipid metabolism disorders that increased levels of TGs, T-CHO and LDL-C in jennies with hyperlipemia. Similar results were presented in the previous findings for donkeys and pony mares [23,25]. Serum TG is commonly used to determine the risk of a donkey becoming hyperlipemic. Donkeys with serum TG concentrations higher than the normal range should begin treatment to correct negative energy balance [26]. Hyperlipemia in donkeys may be due to an increased secretion of VLDLs by the liver, disturbed catabolism of VLDLs and decrease in TGs [27]. This increased LDL-C concentration in the serum was associated with the deficiency of the LDL-C receptor through its failure to function properly [6].

Oxidative stress plays a key role in the pathogenesis of chronic hyperlipemia. The accumulation of high levels of ROS induces oxidative stress, which leads to the damage of DNA, proteins and lipids. In the present study, the lower serum level of ALB and higher serum biochemical parameters related to liver function, including AST, ALT, AKP and CHE, were observed in hyperlipemic jennies. The serum concentration of ALB reflects a balance between ALB synthesis in the liver and its catabolism. ALB also functions as an extracellular antioxidant [28]. The findings of the elevation in the activity of AST, ALT, AKP and CHE in hyperlipemic jennies are in concordance with those reported for hyperlipemic rats [29]. This represents an increase in oxidative factors due to a high serum lipid concentration, which reduces antioxidant defenses and increases lipid peroxidation in the liver.

To date, little has been known about serum trace elements and vitamin change in hyperlipemia in jennies. Evaluating their potential necessity would be valuable. We found significantly lower serum levels of Se in hyperlipemic jennies compared to the control group. Nonetheless, the analysis of serum Se alterations in the published studies on humans with dyslipidemia is highly controversial [30]. Se is an essential trace element that has the ability to counteract free radicals and protect the structure and function of proteins, DNA and chromosomes from oxidative injury [31]. Its biological function is expressed through biologically active compounds, including glutathione peroxidases and other selenoproteins, which have important enzymatic functions associated with antioxidant activity. Furthermore, our results show that the serum level of Se positively correlates with TGs and ALB in hyperlipemic animals, similar to the results previously reported for humans [30,32]. Indeed, Se has been regarded as one of the important factors for lipid metabolism [33]. The evidence supports the idea that Se plays a vital role in insulin mimicry and anti-diabetes [34,35]. Donkeys have innate insulin resistance, especially obese or hyperlipemic jennies [6]. Our present findings indicate an impaired function of selenoproteins and liver metabolism, and the antioxidant capacity of the jennies may have been compromised during late pregnancy. The detailed biological process of selenoproteins involved in lipid metabolism in hyperlipemic jennies needs to be further investigated.

Zinc is known to contribute to the metabolic processes of lipids [17,36]. A previous study found a decrease in serum Zn levels in overweight and obese individuals [37]. However, our study shows no significant differences in serum Zn levels between the two groups. Consistent with our findings, a systematic review confirmed that serum zinc was not irrelevant to dyslipidemia in human studies [17]. The discrepant results may be due to different types of blood lipids and the high intersubject variability in the study population. In addition, our results show that the serum level of Zn is positively associated with the activity of T-AOC in jennies with hyperlipemia. Zn is an integral part of key antioxidant enzymes, and Zn deficiency impairs their synthesis, resulting in increased oxidative stress [38]. A systematic review and meta-analysis showed that Zn supplementation has favorable effects on plasma lipid parameters [39].

An antioxidant defense system, comprising enzymes such as T-SOD and non-enzymatic compounds (e.g., GSH), can prevent oxidative damage to lipoproteins in the blood [40]. In the present study, we observed decreased activities of serum T-SOD and GSH in jennies with hyperlipemia, resulting in a significantly limited antioxidant capacity. The significant decline in GSH may be due to enhanced oxidation or its consumption by electrophilic compounds, such as lipid peroxidation aldehydes [41]. Furthermore, it has been reported that there is a decreased GSH concentration in erythrocytes from hyperlipemic individuals, which has been found to be partly related to high cholesterol levels [42]. Additionally, oxidative stress may produce toxic substances, such as MDA. The formation of MDA is commonly considered a hallmark of cell damage and hepatic injury [40].

Based on the ROC analysis, the AUCs for T-CHO, ALB, AKP, CHE, T-SOD, GSH and MDA were relatively high (0.972, 0.889, 1.000, 0.951, 1.000, 0.917 and 1.000, respectively). These indices may be clinically useful with respect to screening for hyperlipemia in late pregnant jennies. However, serum selenium, zinc and vitamin E could not be used for evaluating the severity of donkey hyperlipemia due to the lower AUC.

5. Conclusions

In summary, significant alterations were observed in various serum lipids, biochemical parameters, antioxidant elements and oxidative stress parameters in jennies with and without hyperlipemia. These changes reflect the metabolic disorders, liver dysfunction and oxidative stress in late pregnant hyperlipemic jennies, providing a basis for the improvement of clinical diagnostic methods and the early prevention and control of hyperlipemia in jennies. The exact effect of antioxidant supplements, such as Se and enzymes, to prevent lipid disorders in late pregnant jennies should be verified in additional specifically designed clinical studies.

Author Contributions: Conceptualization, L.D.; methodology, Q.M., Y.S. and W.L.; software, Q.M., W.L. and X.W.; validation, Q.M., Y.S. and J.L.; formal analysis, Q.M., Y.S., W.L. and J.L.; investigation, Q.M., W.L. and J.L.; data curation, Q.M., Y.S. and W.L.; visualization, Q.M. and X.W.; writing—original draft preparation, Q.M., Y.S. and L.D.; writing—review and editing, L.D.; supervision, L.D. All authors have read and agreed to the published version of the manuscript.

Funding: The program was funded by the Department of Education of Liaoning Province, China (LSNFW201904).

Institutional Review Board Statement: This study was approved by the Animal Care and Use Committee of Shenyang Agricultural University (Approval no. 202306017).

Informed Consent Statement: Informed consent statements were acquired from the donkey owners.

Data Availability Statement: The data presented in this study are available on request from the corresponding authors.

Acknowledgments: We thank Daoxian Sui from the Shenghongdaoda donkey breeding farm, Dalian, China, for the sampling. We also thank Jianbao Dong from the Qingdao Agricultural University for the advice.

Conflicts of Interest: The authors declare no conflicts of interests.

References

1. Ali, M.; Baber, M.; Hussain, T.; Awan, F.; Nadeem, A. The contribution of donkeys to human health. *Equine Vet. J.* **2014**, *46*, 766–767. [CrossRef] [PubMed]
2. Fielding, D. Reproductive characteristics of the jenny donkey—*Equus asinus*: A review. *Trop. Anim. Health Prod.* **1988**, *20*, 161–166. [CrossRef] [PubMed]
3. Frape, D. *Equine Nutrition and Feeding*, 4th ed.; Wiley-Blackwell: Hoboken, NJ, USA, 2010.
4. Martin-Rosset, W. Donkey nutrition and feeding: Nutrient requirements and recommended allowances—A review and prospect. *J. Equine Vet. Sci.* **2018**, *65*, 75–85. [CrossRef]
5. Watson, T. Equine hyperlipemia. In *Metabolic and Endocrine Problems of the Horse*; WB Saunders: London, UK; Britain, UK, 1998; pp. 23–40.
6. Mendoza, F.J.; Toribio, R.E.; Perez-Ecija, A. Metabolic and endocrine disorders in donkeys. *Vet. Clin. N. Am. Equine Pract.* **2019**, *35*, 399–417. [CrossRef] [PubMed]
7. De Lima, B.; Da Silva, G.B.; Da Silva, C.J.F.L.; Ferreira, L.M.C.; Da Costa, C.M.H.E.C.; Filho, H.C.M. Blood, metabolic and endocrine biomarkers in donkeys (*Equus africanus asinus*) supplemented with different energy sources. *Acta Vet. Bras.* **2016**, *10*, 135–143.
8. Agina, O.A. Haematology and clinical biochemistry findings associated with equine diseases—A review. *Not. Sci. Biol.* **2017**, *9*, 1–21. [CrossRef]
9. Harrison, A.; Rickards, K. Hyperlipaemia in donkeys. *UK-Vet Equine* **2018**, *2*, 154–157. [CrossRef]
10. Burden, F.A.; Du Toit, N.; Hazell-Smith, E.; Trawford, A.F. Hyperlipemia in a population of aged donkeys: Description, prevalence, and potential risk factors. *J. Vet. Intern. Med.* **2011**, *25*, 1420–1425. [CrossRef]
11. D'Archivio, M.; Annuzzi, G.; Varì, R.; Filesi, C.; Giacco, R.; Scazzocchio, B.; Santangelo, C.; Giovannini, C.; Rivellese, A.A.; Masella, R. Predominant role of obesity/insulin resistance in oxidative stress development. *Eur. J. Clin. Investig.* **2012**, *42*, 70–78. [CrossRef]
12. Spickett, C.M.; Wiswedel, I.; Siems, W.; Zarkovic, K.; Zarkovic, N. Advances in methods for the determination of biologically relevant lipid peroxidation products. *Free Radic. Res.* **2010**, *44*, 1172–1202. [CrossRef]
13. Yang, R.L.; Shi, Y.H.; Hao, G.; Li, W.; Le, G.W. Increasing oxidative stress with progressive hyperlipidemia in human: Relation between malondialdehyde and atherogenic index. *J. Clin. Biochem. Nutr.* **2008**, *43*, 154–158. [CrossRef] [PubMed]
14. Kiełczykowska, M.; Kocot, J.; Paździor, M.; Musik, I. Selenium—A fascinating antioxidant of protective properties. *Adv. Clin. Exp. Med.* **2018**, *27*, 245–255. [CrossRef] [PubMed]
15. Zago, M.P.; Oteiza, P.I. The antioxidant properties of zinc: Interactions with iron and antioxidants. *Free Radic. Biol. Med.* **2001**, *31*, 266–274. [CrossRef] [PubMed]

16. Niki, E.; Noguchi, N. Dynamics of antioxidant action of vitamin E. *Acc. Chem. Res.* **2004**, *37*, 45–51. [CrossRef] [PubMed]
17. Li, C.P.; Song, Y.X.; Lin, Z.J.; Ma, M.L.; He, L.P. Essential trace elements in patients with dyslipidemia: A meta-analysis. *Curr. Med. Chem.* **2024**, *31*, 3604–3623. [CrossRef]
18. Wong, S.K.; Chin, K.Y.; Suhaimi, F.H.; Ahmad, F.; Ima-Nirwana, S. Vitamin E as a potential interventional treatment for metabolic syndrome: Evidence from animal and human studies. *Front. Pharmacol.* **2017**, *8*, 444. [CrossRef]
19. The Donkey Sanctuary. Hyperlipaemia and the endocrine system. In *The Clinical Companion of the Donkey*; Evans, L., Crane, M., Eds.; Matador Publishing Inc.: Leicestershire, UK, 2018; p. 92.
20. The Donkey Sanctuary. The Donkey Sanctuary. The clinical examination. In *The Clinical Companion of the Donkey*; Evans, L., Crane, M., Eds.; Matador Publishing Inc.: Leicestershire, UK, 2018; p. 256.
21. Liao, Q.; Li, Z.; Han, Y.; Deng, L. Comparative analysis of serum mineral and biochemical parameter profiles between late pregnant and early lactating jennies. *J. Equine Vet. Sci.* **2021**, *99*, 103401. [CrossRef]
22. National Research Council. Donkeys and other equids. In *Nutrient Requirements of Horses*, 6th ed.; The National Academy Press: Washington, DC, USA, 2007.
23. Bonelli, F.; Rota, A.; Corazza, M.; Serio, D.; Sgorbini, M. Hematological and biochemical findings in pregnant, postfoaling, and lactating jennies. *Theriogenology* **2016**, *85*, 1233–1238. [CrossRef]
24. Durham, A.E.; Thiemann, A.K. Nutritional management of hyperlipaemia. *Equine Vet. Educ.* **2015**, *27*, 482–488. [CrossRef]
25. Watson, T.D.; Burns, L.; Packard, C.J.; Shepherd, J. Effects of pregnancy and lactation on plasma lipid and lipoprotein concentrations, lipoprotein composition and post-heparin lipase activities in Shetland pony mares. *J. Reprod. Fertil.* **1993**, *97*, 563–568. [CrossRef]
26. Burden, F.A.; Hazell-Smith, E.; Mulugeta, G.; Patrick, V.; Trawford, R.; Brownlie, H.W.B. Reference intervals for biochemical and haematological parameters in mature domestic donkeys (*Equus asinus*) in the UK. *Equine Vet. Educ.* **2016**, *28*, 134–139. [CrossRef]
27. Watson, T.D.; Murphy, D.; Love, S. Equine hyperlipaemia in the United Kingdom: Clinical features and blood biochemistry of 18 cases. *Vet. Rec.* **1992**, *131*, 48–51. [CrossRef] [PubMed]
28. Yamato, M.; Shiba, T.; Yoshida, M.; Ide, T.; Seri, N.; Kudou, W.; Kinugawa, S.; Tsutsui, H. Fatty acids increase the circulating levels of oxidative stress factors in mice with diet-induced obesity via redox changes of albumin. *FEBS J.* **2007**, *274*, 3855–3863. [CrossRef] [PubMed]
29. Ghosian Moghaddam, M.H.; Roghani, M.; Maleki, M. Effect of hypericum perforatum aqueous extracts on serum lipids, aminotransferases, and lipid peroxidation in hyperlipidemic rats. *Res. Cardiovasc. Med.* **2016**, *5*, e31326.
30. Ju, W.; Ji, M.; Li, X.; Li, Z.; Wu, G.; Fu, X.; Yang, X.; Gao, X. Relationship between higher serum selenium level and adverse blood lipid profile. *Clin. Nutr.* **2018**, *37*, 1512–1517. [CrossRef]
31. Reddy, K.P.; Sailaja, G.; Krishnaiah, C. Protective effects of selenium on fluoride induced alterations in certain enzymes in brain of mice. *J. Environ. Biol.* **2009**, *30* (Suppl. 5), 859–864.
32. Bleys, J.; Navas-Acien, A.; Stranges, S.; Menke, A.; Miller, E.R., 3rd; Guallar, E. Serum selenium and serum lipids in US adults. *Am. J. Clin. Nutr.* **2008**, *88*, 416–423. [CrossRef]
33. Brown, K.M.; Arthur, J.R. Selenium, selenoproteins and human health: A review. *Public Health Nutr.* **2001**, *4*, 593–599. [CrossRef]
34. Stapleton, S.R. Selenium: An insulin mimetic. *Cell Mol. Life Sci.* **2000**, *57*, 1874–1879. [CrossRef]
35. Erbayraktar, Z.; Yılmaz, O.; Artmann, A.T.; Cehreli, R.; Coker, C. Effects of selenium supplementation on antioxidant defense and glucose homeostasis in experimental diabetes mellitus. *Biol. Trace Elem. Res.* **2007**, *118*, 217–226. [CrossRef]
36. Thoen, R.U.; Barther, N.N.; Schemitt, E.; Bona, S.; Fernandes, S.; Coral, G.; Marroni, N.P.; Tovo, C.; Guedes, R.P.; Porawski, M. Zinc supplementation reduces diet-induced obesity and improves insulin sensitivity in rats. *Appl. Physiol. Nutr. Metab.* **2019**, *44*, 580–586. [CrossRef] [PubMed]
37. Rios-Lugo, M.J.; Madrigal-Arellano, C.; Gaytán-Hernández, D.; Hernández-Mendoza, H.; Romero-Guzmán, E.T. Association of serum zinc levels in overweight and obesity. *Biol. Trace Elem. Res.* **2020**, *198*, 51–57. [CrossRef] [PubMed]
38. Kelly, F. Use of antioxidants in the prevention and treatment of disease. *J. Int. Fed. Clin. Chem.* **1998**, *10*, 21–23. [PubMed]
39. Ranasinghe, P.; Wathurapatha, W.S.; Ishara, M.H.; Jayawardana, R.; Galappatthy, P.; Katulanda, P.; Constantine, G.R. Effects of Zinc supplementation on serum lipids: A systematic review and meta-analysis. *Nutr. Metab.* **2015**, *12*, 26. [CrossRef]
40. Zhao, H.; Li, J.; Zhang, J.; Wang, X.; Liu, M.; Zhang, C.; Jia, L. Hepatoprotective and in vitro antioxidant effects of native depolymerised-exopolysaccharides derived from Termitomyces albuminosus. *Sci. Rep.* **2017**, *7*, 3910. [CrossRef]
41. Wolin, M.S. Interactions of oxidants with vascular signaling systems. *Arterioscl. Throm. Vas.* **2000**, *20*, 1430–1442. [CrossRef]
42. Godin, D.V.; Dahlman, D.M. Effects of hypercholesterolemia on tissue antioxidant status in two species differing in susceptibility to atherosclerosis. *Res. Commun. Chem. Pathol. Pharmacol.* **1993**, *79*, 151–166.

veterinary sciences

MDPI

Brief Report

Antimicrobial Resistance Profiling of Pathogens from Cooked Donkey Meat Products in Beijing Area in One Health Context †

Yiting Liu [1,‡], Hongyun Duan [2,‡], Luo Yang [1], Hong Chen [1], Rongzheng Wu [1], Yi Li [1], Yiping Zhu [1,*] and Jing Li [1,3,*]

1 Equine Clinical Diagnostic Centre, College of Veterinary Medicine, China Agricultural University, Beijing 100193, China; sy20233051112@cau.edu.cn (Yiting Liu); lironyoung@cau.edu.cn (L.Y.); sy20233051128@cau.edu.cn (H.C.)
2 Shenzhen Youlan Medical Technology Co., Ltd., Shenzhen 518102, China
3 National Key Laboratory of Veterinary Public Health and Safety, College of Veterinary Medicine, China Agricultural University, Beijing 100193, China
* Correspondence: yipingz@cau.edu.cn (Y.Z.); jlivet@cau.edu.cn (J.L.)
† This article is a revised and expanded version of a paper entitled "Identification and Antimicrobial Resistance Profiling of Pathogens from Donkey Meat Products in Beijing area", which was presented at the 2024 International Forum of Equine Medicine, Wuhan, China, on 24–25 August 2024.
‡ These authors contributed equally to this work.

Simple Summary: Infections from foodborne pathogens can lead to gastroenteritis, which is commonly characterized by diarrhea, abdominal pain, and vomiting. Knowledge of the antimicrobial resistance patterns of foodborne pathogens is crucial for protecting public health. This facilitates the development of strategies to prevent the spread of resistant bacteria and ensures the effectiveness of treatments for foodborne diseases. This study examines the presence and antibiotic resistance of foodborne pathogens in donkey meat products from delis in the Beijing area. The results indicate a potential public health risk due to the presence of antibiotic-resistant bacteria in food products, emphasizing the need for enhanced food safety measures and antibiotic use oversight.

Abstract: The prevalence of foodborne diseases has raised concerns due to the potential transmission of zoonotic bacterial pathogens through meat products. The objective of this study was to determine the occurrence and antimicrobial resistance (AMR) profiles of pathogenic bacteria in cooked donkey meat products from Beijing. Twenty-one cooked donkey meat samples were collected from different delis, subjected to homogenization, and analyzed for bacterial contamination. Molecular identification was performed through polymerase chain reaction (PCR) amplification and sequencing targeting the 16S rDNA gene. The antimicrobial susceptibility of the isolates was evaluated using the disk diffusion method. A total of forty bacterial isolates were identified, with *Proteus mirabilis* being the predominant species, followed by *Klebsiella pneumoniae* and *Novosphingobium*. Both *Proteus mirabilis* and *Klebsiella pneumoniae* exhibited high levels of resistance to several antibiotics, including penicillin, ampicillin, and erythromycin. This study's findings underscore the public health risk posed by antimicrobial-resistant foodborne pathogens and emphasize the necessity for enhanced food safety surveillance within the One Health context.

Keywords: donkey meat product; *Proteus mirabilis*; *Klebsiella pneumoniae*; antimicrobial resistance; One Health

1. Introduction

Foodborne diseases, often attributed to a spectrum of pathogens including bacteria, viruses, fungi, and parasites, pose a significant threat to public health [1]. Bacteria are the predominant etiological agents, causing a myriad of diseases across human and animal populations [1]. Animal-derived meat and meat products, especially those consumed as ready-to-eat (RTE) items, are common vectors for zoonotic bacterial pathogens [2]. The processing of RTE food, which includes cooking, packaging, and storage, is critical in controlling pathogen prevalence throughout the food chain [1]. Nevertheless, the improper handling of these products can lead to contamination and introduce antimicrobial resistance (AMR) bacteria [3]. AMR genes can be transferred to humans through the consumption of meat and meat products containing resistant bacteria, emphasizing the utmost importance of the One Health approach to addressing AMR in public health [4]. Among various bacterial pathogens, *Staphylococcus aureus*, *Salmonella*, and *Listeria monocytogenes* are recognized as major causative agents of foodborne diseases, leading to substantial morbidity and mortality rates worldwide [5].

The increasing prevalence of foodborne disease outbreaks with a microbial etiology in China underscores the urgent need for stringent food safety monitoring [6]. The potential for these bacteria to transfer resistance genes to the human microbiome via the food chain represents a significant public health concern [7]. Donkey meat products are popular in China due to their quality proteins. However, there is a conspicuous lack of comprehensive data on the prevalence of pathogens and the antibiotic resistance phenotype in donkey meat products within China. This study aims to investigate the prevalence of bacterial pathogens and their susceptibility to antibiotics in cooked donkey meat samples from different delis in the Beijing area, which specialize in serving donkey meat products. The findings may contribute to the prevention and control of bacterial infection in donkey meat products and a more comprehensive understanding of food safety concerns from a One Health perspective.

2. Materials and Methods

2.1. Sample Collection

A total of 21 cooked donkey meat samples, comprising donkey burgers and stir-fried donkey meat, were collected via random sampling from 15 delis in different districts with notable sales volumes in the Beijing area. Each sample was aseptically excised using sterile instruments, encompassing both the surface and internal portions of the cooked donkey meat. After collection, the samples were transferred to the laboratory under preservation conditions at 4 °C.

2.2. Isolation and Identification

For the initial homogenization and pre-enrichment, 25 g of each sample was processed in 225 mL of modified tryptone soy broth (Qingdao Haibo, Qingdao, China) and incubated at 37 °C for 18 to 24 h. A 1 mL sample was diluted with 9 mL of sterile brain heart infusion broth (Qingdao Haibo, Qingdao, China) followed by thorough mixing. Subsequently, aliquots were spread onto MacConkey and Columbia blood agar plates (Qingdao Haibo, Qingdao, China) for further incubation at 37 °C for an additional 18 to 24 h to selectively isolate and identify bacterial colonies based on metabolic characteristics and to detect potential pathogens.

2.3. PCR Amplification and Identification

PCR amplification was conducted using universal 16S rDNA primers: the forward primer 27F (5′-AGAGTTTGATCCTGGCTCAG-3′) and the reverse primer 1492R (5′-TACGA CTTAACCCCAATCGC-3′). The reaction mixture comprised 12.5 μL of 2 × Taq PCR Master Mix, 1 μL each of the forward and reverse primers, 0.5 μL of DNA template, and 10 μL of ddH_2O. The thermal cycling protocol consisted of an initial denaturation at 95 °C for 5 min, followed by 30 cycles of 94 °C for 1 min, annealing at 55 to 58 °C for 1 min, and extension at 72 °C for 90 s, with a final extension at 72 °C for 10 min.

After the reaction, 10 μL of the PCR product was loaded on a 1.2% TAE-buffered agarose gel at 120 V for 25 min. The gel was visualized and documented using a UV imaging system with the DL 2000 DNA Marker (TIANGEN BIOTECH (BEIJING) Co., Ltd., Beijing, China) as a size reference. Positive PCR products were submitted for sequencing.

2.4. Antimicrobial Susceptibility Testing

The antibiotic susceptibility of bacterial isolates was assessed using the disk diffusion method, following the guidelines recommended by the Clinical and Laboratory Standards Institute (CLSI). The isolates were tested against a panel of 11 antibiotics: penicillin (PEM), ampicillin (AMP), ceftriaxone (CRO), cefazolin (CZ), tetracycline (TCN), amikacin (AMI), gentamicin (GEN), chloramphenicol (CHL), trimethoprim–sulfamethoxazole (SXT), ciprofloxacin (CIP), norfloxacin (NOR), erythromycin (ERY), and lincomycin (LCM). The selection of antibiotic categories and concentrations was determined in accordance with the "Performance Standards for Antimicrobial Susceptibility Testing; 4th Edition, M100" guidelines as stipulated by the CLSI. To ensure the reliability and reproducibility of the findings, each bacterial isolate was independently tested in triplicate.

3. Results

3.1. Bacteriologic Description

A total of 40 bacterial isolates were obtained from the collected donkey meat samples. The genus *Proteus* showed the highest detection rate (40%), with *Proteus mirabilis* (*P. mirabilis*) (32.5%) being the predominant species (Table 1).

Table 1. Bacteria and proportions of 40 isolates derived from donkey meat samples (n = 21).

Bacteria	Count	Ratio (%)
Proteus mirabilis	13	32.5
Novosphingobium	6	15.0
Klebsiella pneumoniae	5	12.5
Escherichia coli	3	7.5
Proteus vulgaris	2	5.0
Streptococcus lutetiensis	2	5.0
Enterobacter hormaechei	2	5.0
Novosphingobium	1	2.5
Proteus vulgaris	1	2.5
Lysobacter	1	2.5
Enterococcus gallinarum	1	2.5
Macrococcus caseolyticus	1	2.5
Citrobacter freundii	1	2.5
Klebsiella oxytoca	1	2.5

3.2. Antimicrobial Susceptibility Patterns

The isolates of *P. mirabilis* demonstrated a 100% resistance rate to penicillin, tetracycline, and lincomycin. Furthermore, the resistance to erythromycin and ampicillin was observed to be 84.62% and 61.55%, respectively (Table 2). Similarly, *Klebsiella pneumoniae* (*K. pneumoniae*) exhibited complete resistance to penicillin and erythromycin, with an 80% resistance rate to ampicillin (Table 2).

Novosphingobium also presented a 100% resistance profile against penicillin, ampicillin, and erythromycin (Table 3). These findings underscore the prevalence of AMR among the studied bacterial isolates.

Table 2. Antimicrobial susceptibility results for *P. mirabilis* and *K. pneumoniae*.

Antibiotic Category	Antibiotics	*Proteus mirabilis*			*Klebsiella pneumoniae*		
		R (%)	S (%)	Sample Size	R (%)	S (%)	Sample Size
β-lactam antibiotics	PEM	100.0 (13)	0.0 (0)	13	100.0 (5)	0.0 (0)	5
	AMP	61.5 (8)	30.8 (4)	13	80.0 (4)	0.0 (0)	5
	CRO	8.3 (1)	66.7 (8)	12	-	-	-
	CZ	-	-	-	0.0 (0)	75.0 (3)	4
Tetracycline antibiotics	TCN	100.0 (12)	0.0 (0)	12	-	-	-
Aminoglycosides	AMI	-	-	-	0.0 (0)	100.0 (4)	4
	GEN	7.7 (1)	84.6 (11)	13	0.0 (0)	100.0 (5)	5
Chloramphenicols	CHL	46.2 (6)	38.5 (5)	13	0.0 (0)	100.0 (5)	5
Sulfonamides	SXT	38.6 (5)	46.2 (6)	13	0.0 (0)	80.0 (4)	5
Fluoroquinolones	CIP	7.7 (1)	84.6 (11)	13	0.0 (0)	100.0 (5)	5
	NOR	-	-	-	0.0 (0)	100.0 (4)	4
Macrolide antibiotics	ERY	84.6 (11)	15.4 (2)	13	100.0 (5)	0.0 (0)	5
Lincosamides	LCM	100.0 (12)	0.0 (0)	12	-	-	-

R (%) = (Number of strains resistant to the antibiotic/Sample size) × 100%. S (%) = (Number of strains sensitive to the antibiotic/Sample size) × 100%. AMK, amikacin; AMP, ampicillin; CHL, chloramphenicol; CIP, ciprofloxacin; CRO, ceftriaxone; CZ, cefazolin; ERY, erythromycin; GEN, gentamicin; LCM, lincomycin; NOR, norfloxacin; PEM, penicillin; SXT, trimethoprim–sulfamethoxazole; TCN, tetracycline.

Table 3. Antimicrobial susceptibility results for *Novosphingobium*.

Antibiotic Category	Antibiotics	*Novosphingobium*		
		R (%)	S (%)	Sample Size
β-lactam antibiotics	PEM	100.0 (6)	0.0 (0)	6
	AMP	100.0 (6)	0.0 (0)	6
	CZ	0.0 (0)	83.3 (5)	6
Aminoglycosides	AMK	0.0 (0)	100.0 (6)	6
	GEN	0.0 (0)	100.0 (6)	6
Chloramphenicols	CHL	0.0 (0)	100.0 (6)	6
Sulfonamides	SXT	50.0 (3)	50.0 (3)	6
Fluoroquinolones	CIP	0.0 (0)	100.0 (6)	6
	NOR	0.0 (0)	100.0 (6)	6
Macrolide antibiotics	ERY	100.0 (6)	0.0 (0)	6

4. Discussion

This study identified the presence of AMR bacteria in cooked donkey meat products from delis in the Beijing area. The bacterial isolates were predominantly *P. mirabilis*, followed by *K. pneumoniae* and *Novosphingobium*.

P. mirabilis, an opportunistic pathogen, is a well-known cause of nosocomial infections. Additionally, it is frequently implicated in urinary tract infections [8]. Its presence in donkey meat products, as identified in this study, raises concerns due to its potential to cause foodborne illness. *P. mirabilis* is known to inhabit various environmental niches, including soil and aquatic ecosystems, and is often found as a commensal organism in the gastrointestinal tracts of humans and animals [9]. *Proteus* species and *K. pneumoniae* have been isolated as thermotolerant bacteria from equids in India, which may survive in food-processing units and contaminate pasteurized products [10]. In a study by Liu Wei et al., from 2010 to 2019, there were 143 outbreaks of foodborne diseases in Haidian District, Beijing, with a total of 272 strains of pathogens detected, among which *Proteus* accounted for as high as 25.7%, becoming one of the main etiological agents of foodborne disease outbreaks in the area [11]. In a study conducted by Xiaobing et al., a total of 52 isolates of *P. mirabilis* were isolated from 178 cooked meat samples, with an isolation rate of 29.2% [12]. The presence of *P. mirabilis* in retail meat products could represent a pathway for the transmission of this pathogen to humans [13].

There was a marked increase in antibiotic-resistant *Proteus* spp. from 48.4% in 2011 to 74% in 2020 in hospital settings in Italy [14]. A particularly high resistance to macrolides, tetracyclines, fusidic acid, and mupirocin and a 77.6% resistance rate to aminopenicillins in *Proteus* spp. were detected in this research [14]. In the present study, a lower rate of 61.5% resistance to ampicillin was reported in the *P. mirabilis* isolates sourced from cooked donkey meat samples. This variance is probably due to the differing sources of the bacteria. Hospital-acquired strains are more frequently exposed to antibiotics, which likely contributes to increased resistance [14]. This highlights the importance of monitoring the prevalence and resistance patterns of this bacterium in food products from a One Health perspective. In research led by Delphine Girlich et al., *P. mirabilis* was found to be inherently resistant to several antibiotic classes including penicillin, macrolides, tetracyclines, and lincosamides [15]. This observation is consistent with the AMR patterns observed in the current study of *P. mirabilis*. This resistance is concerning as it restricts treatment options for infections caused by this bacterium. These findings underscore the alarming rise in antibiotic resistance among *P. mirabilis* strains, which not only poses a significant challenge to public health but also necessitates a comprehensive approach to antibiotic stewardship in both clinical and food production settings.

K. pneumoniae, recognized for its ability to inhabit the gastrointestinal tract of animals, has been identified as a common contaminant in various retail food products, including meats and vegetables [16]. This bacterium is a potential source of foodborne illness and can lead to extraintestinal infections in humans, posing a considerable public health concern [17]. In a study by Haryani et al., 32% of 78 street food samples from Malaysia were positive for *K. pneumoniae*, with isolates showing 100% resistance to ampicillin, erythromycin, and rifampicin and 80% to sulfamethoxazole [18]. Similarly, another study found *K. pneumoniae* in 7.5% (35/464) of cooked food samples, with the highest resistance noted for ampicillin at 92.3%, followed by tetracycline at 31.3% [19]. Additionally, the study indicated that cooking processes may not always be effective in eliminating bacteria [19]. The presence of *K. pneumoniae* in ready-to-eat cooked foods may suggest suboptimal hygienic practices during food handling or contamination after cooking. Possible sources of contamination include cross-contamination with raw products, meat juices, or other contaminated items or from handlers with inadequate personal hygiene. The thermotolerance of *K. pneumoniae* raises concerns, as it can survive in cooked foods, and its complete inactivation is not always guaranteed, even at temperatures up to 60 °C [10]. The resistance patterns also emphasize the importance of prudent antibiotic use and the implementation of stringent hygiene and handling protocols to prevent the spread of resistant strains.

The genus *Novosphingobium* encompasses a group of Gram-negative bacteria within the class Alphaproteobacteria, which was previously classified within the genus *Sphingomonas* [20]. Characterized by their metabolic versatility, these bacteria are known for their potential roles in the biodegradation of a spectrum of organic compounds [21]. In an investigation led by P. Papademas et al., high-throughput 16S rDNA sequencing revealed that *Sphingomonas* was the predominant genus in donkey milk from a farm in Cyprus, with a notable presence ranging from 17% to 49% [22]. Interestingly, while *Sphingomonas* was not detected in our analysis of donkey meat, we identified *Novosphingobium*, with analogous metabolic capabilities [22]. *Novosphingobium*'s metabolic diversity and role in the biodegradation of organic compounds are well documented, as is its isolation from multiple environments such as soil, water sediments, activated sludge, and even within plant tissues [23,24]. Importantly, this is the first report of *Novosphingobium* isolated from donkey meat, emphasizing the need for further investigation into its ecological distribution and potential impact on food safety. In one study, it was observed that colorectal cancer patients with reduced tumor tissue levels of *Caulobacter* and *Novosphingobium* showed enhanced long-term survival [25]. This may indicate the potential negative impact of *Novosphingobium* on human health. Regarding antibiotic resistance, Belmok et al. reported that *Novosphingobium terrae* exhibited resistance to penicillin, trimethoprim–sulfamethoxazole, and erythromycin [25], which is partially consistent with our findings. The ability of

Novosphingobium to form biofilms and survive in chlorinated conditions raises concerns about its potential to harbor and protect other pathogens, including bacteria, which may contribute to the spread of antibiotic resistance [26]. It is worth noting that research on the antibiotic resistance of *Novosphingobium* is still limited, and further studies are required to fully understand the scope and implications of this issue. In addition, the impact of *Novosphingobium* on food safety is still uncertain.

The contamination of cooked meat products with pathogenic bacteria is a multifaceted issue, influenced by the introduction of bacteria during preparation, the sanitary conditions in the sampling site, and the prevalence of antibiotic resistance [5]. One of the primary factors identified is the presence of pathogenic bacteria such as *Proteus mirabilis, Klebsiella pneumoniae,* and *Escherichia coli*, which are known to thrive in environments lacking proper sanitation [5]. Furthermore, the risk of cooked meat contamination is increased by the conditions in the sampling site. Factors such as inadequate temperature control, the poor personal hygiene of food handlers, and unsanitary equipment can significantly contribute to the proliferation of these bacteria [27]. Our study indicates a high prevalence of *Proteus mirabilis* (32.5%) and *Novosphingobium* (15.0%), suggesting that the sampling site may have been compromised by poor hygiene practices. Additionally, the use of antibiotics in livestock can lead to AMR bacteria, which may contaminate meat during processing. Addressing these factors requires a multifaceted approach that includes stringent sanitation protocols, proper food handling education, and responsible antibiotic stewardship in livestock.

5. Conclusions

This study identified *Proteus mirabilis*, *Klebsiella pneumoniae*, and *Novosphingobium* as the dominant bacteria in cooked donkey meat products in the Beijing area, with notable resistance to common antibiotics including penicillin, ampicillin, and erythromycin. The findings underscore the need for further AMR monitoring and control in meat chains within the One Health context.

Author Contributions: Conceptualization, J.L. and Y.Z.; Methodology, Y.L. (Yiting Liu) and H.D.; Software, L.Y.; Validation, H.C., R.W. and L.Y.; Formal Analysis, Y.L. (Yiting Liu); Investigation, H.D.; Resources, J.L.; Data Curation, H.C.; Writing—Original Draft Preparation, Y.L. (Yiting Liu); Writing—Review & Editing, Y.Z., H.D. and Y.L. (Yi Li); Visualization, R.W.; Supervision, J.L.; Project Administration, Y.Z.; Funding Acquisition, J.L. All authors have read and agreed to the published version of the manuscript.

Funding: This research was funded by National Natural Science Foundation of China, grant number 32202861.

Institutional Review Board Statement: Not applicable.

Informed Consent Statement: Not applicable.

Data Availability Statement: Data are contained within the article.

Conflicts of Interest: Author Hongyun Duan was employed by the company Shenzhen Youlan Medical Technology Co., Ltd. The remaining authors declare that the research was conducted in the absence of any commercial or financial relationships that could be construed as a potential conflict of interest.

References

1. Abalkhail, A. Frequency and Antimicrobial Resistance Patterns of Foodborne Pathogens in Ready-to-Eat Foods: An Evolving Public Health Challenge. *Appl. Sci.* **2023**, *13*, 12846. [CrossRef]
2. Ali, S.; Alsayeqh, A.F. Review of Major Meat-Borne Zoonotic Bacterial Pathogens. *Front. Public Health* **2022**, *10*, 1045599. [CrossRef] [PubMed]
3. Conceição, S.; Queiroga, M.C.; Laranjo, M. Antimicrobial Resistance in Bacteria from Meat and Meat Products: A One Health Perspective. *Microorganisms* **2023**, *11*, 2581. [CrossRef] [PubMed]
4. Nastasijevic, I.; Proscia, F.; Jurica, K.; Veskovic-Moracanin, S. Tracking Antimicrobial Resistance Along the Meat Chain: One Health Context. *Food Rev. Int.* **2024**, *40*, 2775–2809. [CrossRef]

5. Rajaei, M.; Moosavy, M.H.; Gharajalar, S.N.; Khatibi, S.A. Antibiotic Resistance in the Pathogenic Foodborne Bacteria Isolated from Raw Kebab and Hamburger: Phenotypic and Genotypic Study. *BMC Microbiol.* **2021**, *21*, 272. [CrossRef] [PubMed]
6. Wu, Y.; Liu, X.; Chen, Q.; Liu, H.; Dai, Y.; Zhou, Y.J.; Wen, J.; Tang, Z.Z.; Chen, Y. Surveillance for Foodborne Disease Outbreaks in China, 2003 to 2008. *Food.Cont.* **2018**, *84*, 382–388. [CrossRef]
7. Cao, H.; Bougouffa, S.; Park, T.J.; Lau, A.; Tong, M.K.; Chow, K.H.; Ho, P.L. Sharing of Antimicrobial Resistance Genes between Humans and Food Animals. *MSystems* **2022**, *7*, e00775-22. [CrossRef]
8. Mark, D.G.; Hung, Y.Y.; Salim, Z.; Tarlton, N.J.; Torres, E.; Frazee, B.W. Third-Generation Cephalosporin Resistance and Associated Discordant Antibiotic Treatment in Emergency Department Febrile Urinary Tract Infections. *Ann. Emerg. Med.* **2021**, *78*, 357–369. [CrossRef]
9. Girlich, D.; Bonnin, R.A.; Dortet, L.; Naas, T. Genetics of Acquired Antibiotic Resistance Genes in *Proteus* spp. *Front. Microbiol.* **2020**, *11*, 256. [CrossRef]
10. Singh, B.R. Thermotolerance and multidrug resistance in bacteria isolated from equids and their environment. *Vet. Rec.* **2009**, *164*, 746–750. [CrossRef]
11. Liu, W.; Yin, K.; Shi, L.; Han, X.; Bai, J.; Ji, L.; Han, S. Analysis of the Results of Foodborne Disease Outbreaks in Haidian District, Beijing, from 2010 to 2019. *Strait J. Prev. Med.* **2020**, *28*, 91–93. (In Chinese)
12. Jiang, X.; Yu, T.; Liu, L.; Li, Y.; Zhang, K.; Wang, H.; Shi, L. Examination of Quaternary Ammonium Compound Resistance in *Proteus mirabilis* Isolated from Cooked Meat Products in China. *Front. Microbiol.* **2017**, *8*, 2417. [CrossRef] [PubMed]
13. Ma, W.; Han, Y.; Zhou, L.; Peng, W.; Mao, L.; Yang, X.; Wang, Q.; Zhang, T.; Wang, H.; Lei, C. Contamination of *Proteus mirabilis* Harbouring Various Clinically Important Antimicrobial Resistance Genes in Retail Meat and Aquatic Products from Food Markets in China. *Front. Microbiol.* **2022**, *13*, 1086800. [CrossRef] [PubMed]
14. Facciolà, A.; Gioffrè, M.E.; Chiera, D.; Ferlazzo, M.; Virga, A.; Laganà, P. Evaluation of Antibiotic Resistance in *Proteus* spp: A Growing Trend that Worries Public Health. Results of 10 Years of Analysis. *New Microbiol.* **2022**, *45*, 269–277. [PubMed]
15. Stock, I. Natural Antibiotic Susceptibility of *Proteus* spp., with Special Reference to *P. mirabilis* and *P. penneri* Strains. *J. Chemother.* **2003**, *15*, 12–26. [CrossRef]
16. Crippa, C.; Pasquali, F.; Rodrigues, C.; De, C.A.; Lucchi, A.; Gambi, L.; Manfreda, G.; Brisse, S.; Palma, F. Genomic Features of *Klebsiella* Isolates from Artisanal Ready-to-eat Food Production Facilities. *Sci. Rep.* **2023**, *13*, 10957. [CrossRef]
17. Hartantyo, S.H.P.; Chau, M.L.; Koh, T.H.; Yap, M.; Yi, T.; Cao, D.Y.H.; GutiÉrrez, R.A.; Ng, L.C. Foodborne *Klebsiella pneumoniae*: Virulence Potential, Antibiotic Resistance, and Risks to Food Safety. *J. Food Prot.* **2020**, *83*, 1096–1103. [CrossRef]
18. Haryani, Y.; Noorzaleha, A.S.; Fatimah, A.B.; Noorjahan, B.A.; Patrick, G.B.; Shamsinar, A.T.; Laila, R.A.S.; Son, R. Incidence of *Klebsiella pneumonia* in street foods sold in Malaysia and their characterization by antibiotic resistance, plasmid profiling, and RAPD–PCR analysis. *Food Cont.* **2007**, *18*, 847–853. [CrossRef]
19. Guo, Y.; Zhou, H.; Qin, L.; Pang, Z.; Qin, T.; Ren, H.; Pan, Z.; Zhou, J. Frequency, Antimicrobial Resistance and Genetic Diversity of *Klebsiella pneumoniae* in Food Samples. *PLoS ONE* **2016**, *11*, e0153561. [CrossRef]
20. Liu, Y.; Pei, T.; Du, J.; Huang, H.; Deng, M.; Zhu, H. Comparative Genomic Analysis of the Genus *Novosphingobium* and the Description of Two Novel Species *Novosphingobium aerophilum* sp. nov. and *Novosphingobium jiangmenense* sp. nov. *Syst. Appl. Microbiol.* **2021**, *44*, 126202. [CrossRef]
21. Goswami, K.; Deka, B.H.P.; Saikia, R. Purification and Characterization of Cellulase Produced by *Novosphingobium* sp. Cm1 and its Waste Hydrolysis Efficiency and Bio-stoning Potential. *J. Appl. Microbiol.* **2022**, *132*, 3618–3628. [CrossRef] [PubMed]
22. Papademas, P.; Kamilari, E.; Aspri, M.; Anagnostopoulos, D.A.; Mousikos, P.; Kamilaris, A.; Tsaltas, D. Investigation of Donkey Milk Bacterial Diversity by 16S rDNA High-throughput Sequencing on a Cyprus Donkey Farm. *J. Dairy Sci.* **2021**, *104*, 167–178. [CrossRef] [PubMed]
23. Zhang, X.; Liu, Y.; Lin, Y.; Wang, L.; Yao, S.; Cao, Y.; Zhai, L.; Tang, X.; Zhang, L.; Zhang, T.; et al. *Novosphingobium clariflavum* sp. nov., Isolated from a Household Product Plant. *Int. J. Syst. Evol. Microbiol.* **2017**, *67*, 3150–3155. [CrossRef] [PubMed]
24. Rangjaroen, C.; Sungthong, R.; Rerkasem, B.; Teaumroong, N.; Noisangiam, R.; Lumyong, S. Untapped Endophytic Colonization and Plant Growth-Promoting Potential of the Genus *Novosphingobium* to Optimize Rice Cultivation. *Microbes Environ.* **2017**, *32*, 84–87. [CrossRef]
25. Zhou, B.; Shi, L.; Jin, M.; Cheng, M.; Yu, D.; Zhao, L.; Zhang, J.; Chang, Y.; Zhang, T.; Liu, H. Caulobacter and *Novosphingobium* in Tumor Tissues are Associated with Colorectal Cancer Outcomes. *Front. Oncol.* **2023**, *12*, 1078296. [CrossRef]
26. Narciso-da-Rocha, C.; Vaz-Moreira, I.; Manaia, C.M. Genotypic Diversity and Antibiotic Resistance in *Sphingomonadaceae* Isolated from Hospital Tap Water. *Sci. Total Environ.* **2014**, *466–467*, 127–135. [CrossRef]
27. Park, M.; Wang, J.; Park, J.; Forghani, F.; Moon, J.; Oh, D. Analysis of Microbiological Contamination in Mixed Pressed Ham and Cooked Sausage in Korea. *J. Food Prot.* **2014**, *77*, 412–418. [CrossRef]

MDPI AG
Grosspeteranlage 5
4052 Basel
Switzerland
Tel.: +41 61 683 77 34

Veterinary Sciences Editorial Office
E-mail: vetsci@mdpi.com
www.mdpi.com/journal/vetsci

www.ingramcontent.com/pod-product-compliance
Lightning Source LLC
Chambersburg PA
CBHW062053220926
51850CB00010B/2758